D0215197

Fundamentals of Modern Elementary Geometry

All that a man has to say or do that can possibly concern mankind is, in some shape or other, to tell the story of his love . . .

HENRY DAVID THOREAU

FUNDAMENTALS OF MODERN ELEMENTARY GEOMETRY

HOWARD EVES

Distinguished Visiting Professor
University of Central Florida

JONES AND BARTLETT PUBLISHERS

Boston *London*

Editorial, Sales, and Customer Service Offices
Jones and Bartlett Publishers
20 Park Plaza
Boston, MA 02116

Jones and Bartlett Publishers International
P.O. Box 1498
London W6 7RS
England

Library of Congress Cataloging-in-Publication Data

Eves, Howard Whitley, 1911–
Fundamentals of modern elementary geometry / Howard Eves.
p. cm.
Includes bibliographical references and index.
ISBN 0-86720-247-5
1. Geometry. I. Title.
QA453.E94 1992
516′.04--dc20 91-42971
CIP

Printed in the United States of America

96 95 94 93 92 10 9 8 7 6 5 4 3 2 1

TO DIANE

We started rambling through the hills together
Then we decided to ramble through life together

Contents

Foreword

To prepare oneself to teach high school geometry, one should do two things: (1) learn considerably more elementary geometry than is covered in the high school course itself and (2) become acquainted with modern notions of geometric structure.

The first of the above requirements follows from the well-known pedagogical fact that one cannot give a strong, interesting, and inspiring course in a subject without knowing a good deal more about the subject than one teaches. The second of the two requirements follows from the fact that one cannot attain a true appreciation of present-day geometry without knowing something of the evolution of geometric structure from the days of the ancient Greeks up into modern times.

The above observations led to the decision to write one-semester college texts, one devoted to modern elementary geometry and another to the evolution of geometric structure. The first text is a sequel to basic high school geometry and introduces the reader to some of the important modern extensions of elementary geometry—extensions that have largely entered into the mainstream of mathematics. The second text will treat notions of geometric structure that arose with the non-Euclidean revolution in the first half of the ninteenth century. It is hoped that these two texts will furnish a suitable introduction to (so-called) *college geometry* on the one hand and *non-Euclidean geometry* on the other. With the help of these two texts, teachers should be better able to rescue high school geometry from the doldrums, reviving some of its inherent romance, beauty, and excitement.

Herewith is the first of the two texts.

Preface

A number of people have requested that a concise, one-semester, first-course text on modern elementary geometry, aimed largely at prospective and current teachers of high school geometry, be constructed from Volume One of my more massive *A Survey of Geometry* (Boston: Allyn and Bacon, Inc., 1963, revised in 1972, and now out of print). The present work is an attempt to satisfy these requests.

Here, briefly, are some of the features of the work.

1. *Geometric content.* It is obvious that a teacher of high school geometry should know appreciably more about elementary geometry than just the material in the high school texts. The additional information should be a sequel to, or extension of, that which is taught in high school, and it should not (at least in a first course) run too far afield from the subject matter of high school courses. Much of the material should be such that it can be passed on to a gifted or interested geometry student, and it should tend to knit together parts of the usual high school information. Also, in general, it is wise to introduce material that has, in one form or another, entered into the mainstream of mathematics. These are the guidelines that were followed in selecting the geometric content of this book.

2. *Geometric concepts.* The two most significant general concepts that can be introduced in a course in elementary geometry are the idea of a deductive chain of statements and the idea of geometric transformation. Each section of this work is essentially an illustration of a deductive development of some important and connected part of geometry, and the transform-solve-invert technique of transformation theory constitutes the red thread that runs through the entire work. The former concept leads to the axiomatic foundation of geometry, and the latter leads to the group-theoretic foundation of geometry. In a later course, these two concepts can be more sharply pursued—introducing the student on the one hand to non-Euclidean geometry, axiomatics, and finite and abstract geometries and on the other hand to the Erlanger Programm and Klein's remarkable codification of geometries.

In addition to exposing the student to these two principal concepts, the text explores a number of lesser concepts, such as directed elements, ideal elements, cross ratio, and duality.

3. *Rich geometrical attack*. The old saying "There is more than one way to skin a cat" is certainly true in geometry; there is no unique way to prove a theorem or to solve a problem. To illustrate this fact, several important theorems have been established in a number of different ways as suitable methods have been introduced. Thus some key theorems, such as the Desargues two-triangle theorem and the Pascal mystic-hexagram theorem, appear and reappear as variations on a geometric theme.

4. *Underlying history*. It has quite properly been said that "no subject loses more than does mathematics when dissociated from its history." The treatment in this book is strongly historical, for the study is concerned with fundamental ideas, and a genuine appreciation and understanding of ideas is not possible without an analysis of their origins. The history of a subject enlivens it, heightens its appeal, and brings out valuable cultural links.

5. *Flexibility*. A certain flexibility is desirable in a text—so that the text can fit classes with different degrees of preparation, courses of various lengths, and the individual aims of particular instructors. The present text can be used in a number of ways. A well-prepared class may cover all three chapters of the book in a semester, probably omitting the six sections marked optional. A less well prepared class or a class geared to a trimester system might restrict itself to Chapter 1, Chapter 2, and the first three sections of Chapter 3, or even to just Chapters 1 and 2, in either case omitting the optional sections. In any of the above selections, if more time is needed for the remaining material, Sections 2.9 and 2.10 can also be omitted. Another possibility is merely to start at the beginning and proceed only as far as time and pace permit. Which arrangement an instructor follows will also depend on how much time is devoted to the problem material. The optional sections make excellent assigned reading for the especially bright students in a class.

6. *Problems*. It is difficult to overstate the importance of problems in mathematics in general and geometry in particular. Problems are an integral part of geometry, for one learns geometry chiefly by doing geometry. This text contains an ample supply of problems, many of which introduce the student to interesting matter not appearing in the text proper. The student will miss much if he or she does not dip into the problem material. Since geometry problems often present difficulties to a beginner, suggestions for solutions of most of the problems are given at the end of the work.

7. *Bibliography*. At the end of each chapter is a short bibliography of related works.

1

Modern Elementary Geometry

The period following the European Renaissance, and running into present times, is known in the history of mathematics as *the modern era*. One of the ways mathematicians of the modern era have extended geometry beyond that inherited from the ancient Greeks has been through the discovery of a host of further properties of circles and rectilinear figures, deduced from those listed in the early books of Euclid's *Elements* (written about 300 B.C.). This material is referred to as *modern elementary geometry* and is a sequel to or expansion of high school geometry, much of which is based on Euclid's *Elements*.

The nineteenth century witnessed an astonishing growth in modern elementary geometry, and the number of papers on the subject that have since appeared is almost unbelievable. In 1906, Maximilian Simon (in his *Über die Entwickelung der Elementargeometrie im XIX Jahrhundert*) attempted the construction of a catalogue of contributions to elementary geometry made during the nineteenth century; it has been estimated that this catalogue contains upward of 10,000 references! Research in the field has not abated, and it would seem that the geometry of the triangle and its associated points, lines, and circles must be inexhaustible. Much of the material has been extended to the tetrahedron and its associated points, lines, planes, and spheres, resulting in an enormous and beautiful expansion of elementary solid geometry. Many of the special points, lines, circles, planes, and spheres have been named after original and subsequent investigators. Among these investigators are Gergonne, Nagel, Feuerbach, Hart, Casey, Brocard, Lemoine, Tucker, Neuberg, Simson, McCay, Euler, Gauss, Bodenmiller, Fuhrmann, Schoute, Spieker, Taylor, Droz-Farny, Morley, Miquel, Hagge, Peaucellier, Steiner, and Tarry.

Here is a singularly fascinating, highly intricate, extensive, and challenging field of mathematics that can be studied and pursued with very little mathematical prerequisite and with a fair chance of making an original discovery. The subject, though elementary, is often far from easy. Large portions of the material have been summarized and organized into textbooks bearing the title of modern, or college, geometry.

In this chapter we shall briefly consider a few selected topics from modern elementary geometry. Though we cannot hope in such a short space to do justice to the subject, the topics we choose will illustrate some of the flavor and attractiveness of this area of geometry, and all of them will be useful in later parts of our work. Much of modern elementary geometry has, in one way or another, entered into the mainstream of mathematics.

In Chapters 2 and 4, certain additional topics belonging to modern elementary geometry will be encountered.

1.1 SENSED MAGNITUDES

One of the innovations of modern elementary geometry is the employment, when it proves useful, of sensed, or signed, magnitudes. It was the extension of the number system to include both positive and negative numbers that led to this forward step in geometry. Although Albert Girard, René Descartes, and others introduced negative segments into geometry during the seventeenth century, the idea of sensed magnitudes was first systematically exploited in the early nineteenth century by L. N. M. Carnot (in his *Géométrie de position* of 1803) and especially by A. F. Möbius (in his *Der barycentrische Calcul* of 1827). By means of the concept of sensed magnitudes, several separate statements or relations can often be combined into a single embracive statement or relation, and a single proof can frequently be formulated for a theorem that would otherwise require the treatment of a number of different cases.

We start a study of sensed magnitudes with some definitions and a notation.

1.1.1 DEFINITIONS AND NOTATION. Sometimes we shall choose one direction along a given straight line as the positive direction, and the other direction as the negative direction. A segment AB on the line will then be considered *positive* or *negative* according as the direction from A to B is the positive or negative direction of the line, and the symbol $\overline{AB}$ (in contrast to AB) will be used to denote the resulting signed distance from the point A to the point B. Such a segment $\overline{AB}$ is called a *sensed*, or *directed*, *segment*; point A is called the *initial point* of the segment and point B is called the *terminal point* of the segment. The fact that $\overline{AB}$ and $\overline{BA}$ are equal in magnitude but opposite in direction is indicated by the equation

$$\overline{AB} = -\overline{BA},$$

or by the equivalent equation

$$\overline{AB} + \overline{BA} = 0.$$

Of course $\overline{AA} = 0$.

1.1.2 Definitions. Points that lie on the same straight line are said to be *collinear*. A set of collinear points is said to constitute a *range* of points, and the straight line on which they lie is called the *base* of the range. A range consisting of all the points of its base is called a *complete range*.

We are now in a position to establish a few basic theorems about sensed line segments.

1.1.3 Theorem. *If A, B, C are any three collinear points, then*

$$\overline{AB} + \overline{BC} + \overline{CA} = 0.$$

If the points A, B, C are distinct, then C must lie between A and B, or on the prolongation of $\overline{AB}$, or on the prolongation of $\overline{BA}$. We consider these three cases in turn.

If C lies between A and B, then $\overline{AB} = \overline{AC} + \overline{CB}$, or $\overline{AB} - \overline{CB} - \overline{AC} = 0$, or $\overline{AB} + \overline{BC} + \overline{CA} = 0$.

If C lies on the prolongation of $\overline{AB}$, then $\overline{AB} + \overline{BC} = \overline{AC}$, or $\overline{AB} + \overline{BC} - \overline{AC} = 0$, or $\overline{AB} + \overline{BC} + \overline{CA} = 0$.

If C lies on the prolongation of $\overline{BA}$, then $\overline{CA} + \overline{AB} = \overline{CB}$, or $\overline{AB} - \overline{CB} + \overline{CA} = 0$, or $\overline{AB} + \overline{BC} + \overline{CA} = 0$.

The situations where one or more of the points A, B, C coincide are easily disposed of.

This theorem illustrates one of the economy features of sensed magnitudes. Without the concept of directed line segments, three separate equations would have to be given to describe the possible relations connecting the three unsigned distances AB, BC, CA between pairs of the three distinct collinear points A, B, C.

1.1.4 THEOREM. *Let O be any point on the line of segment AB. Then* $\overline{AB} = \overline{OB} - \overline{OA}$.

This is an immediate consequence of Theorem 1.1.3, for by that theorem we have $\overline{AB} + \overline{BO} + \overline{OA} = 0$, whence $\overline{AB} = -\overline{BO} - \overline{OA} = \overline{OB} - \overline{OA}$.

1.1.5 EULER'S THEOREM (1747). *If A, B, C, D are any four collinear points, then*

$$\overline{AD} \cdot \overline{BC} + \overline{BD} \cdot \overline{CA} + \overline{CD} \cdot \overline{AB} = 0.$$

The theorem follows by noting that, by Theorem 1.1.4, the left member of the above equation may be put in the form

$$\overline{AD}(\overline{DC} - \overline{DB}) + \overline{BD}(\overline{DA} - \overline{DC}) + \overline{CD}(\overline{DB} - \overline{DA}),$$

which, upon expansion and reduction, is found to vanish identically.

The notion of directed segment leads to the following very useful definition of the ratio in which a point on a line divides a segment on that line.

1.1.6 DEFINITIONS. If A, B, P are distinct collinear points, we define *the ratio in which P divides the segment $\overline{AB}$* to be the ratio $\overline{AP}/\overline{PB}$. Note that the value of this ratio is independent of any direction assigned to line AB. If P lies between A and B, the division is said to be *internal*; otherwise the division is said to be *external*. Denoting the ratio $\overline{AP}/\overline{PB}$ by r, we note that if P lies on the prolongation of $\overline{BA}$, then $-1 < r < 0$; if P lies between A and B, then $0 < r < \infty$; if P lies on the prolongation of $\overline{AB}$, then $-\infty < r < -1$.

If A and B are distinct and P coincides with A, we set $\overline{AP}/\overline{PB} = 0$. If A and B are distinct and P coincides with B, the ratio $\overline{AP}/\overline{PB}$ is undefined and we indicate this by writing $\overline{AP}/\overline{PB} = \infty$ or $-\infty$, according as P is thought of as having approached B from the direction of A or not.

Workers in modern elementary geometry have devised several ways of assigning a sense to angles lying in a common plane, and each way has its own uses. The way we are about to describe is particularly useful in relations involving trigonometric functions.

1.1.7 DEFINITIONS AND NOTATION. We may consider an $\sphericalangle AOB$ as generated by the rotation of side OA about point O until it coincides with side OB, the rotation not exceeding 180°. If the rotation is counterclockwise the angle is said to be *positive*; if the rotation is clockwise the angle is said to be *negative*, and the symbol $\sphericalangle \overline{AOB}$ (in contrast to $\sphericalangle AOB$) will be used to denote the resulting signed rotation. Such an $\sphericalangle \overline{AOB}$ is called a *sensed*, or *directed*, *angle*; point O is called the *vertex* of the angle; side OA is called the *initial side* of the angle; side OB is called the *terminal side* of the angle. If $\sphericalangle AOB$ is not a straight angle, then the fact that $\sphericalangle\overline{AOB}$ and $\sphericalangle\overline{BOA}$ are equal in magnitude but opposite in direction is indicated by the equation

$$\sphericalangle\overline{AOB} = -\sphericalangle\overline{BOA},$$

or by the equivalent equation

$$\sphericalangle\overline{AOB} + \sphericalangle\overline{BOA} = 0.$$

It is sometimes convenient to assign a sense to the areas of triangles lying in a common plane.

1.1.8 DEFINITIONS AND NOTATION. A triangle ABC will be considered as *positive* or *negative* according as the tracing of the perimeter from A to B to C to A is counterclockwise or clockwise. Such a signed triangular area is called a *sensed*, or *directed*, *area*, and will be denoted by $\triangle\overline{ABC}$ (in contrast to $\triangle ABC$).

Some subsequent developments in this chapter will make use of the following two important theorems.

1.1.9 THEOREM. *If vertex A of triangle ABC is joined to any point L on line BC, then*

$$\frac{\overline{BL}}{\overline{LC}} = \frac{AB \sin \overline{BAL}}{AC \sin \overline{LAC}}.$$

Let h denote the length of the perpendicular from A to line BC. The reader may then check that for all possible figures,

$$\frac{\overline{BL}}{\overline{LC}} = \frac{h\overline{BL}}{h\overline{LC}} = \frac{2\triangle\overline{ABL}}{2\triangle\overline{ALC}} = \frac{(AB)(AL)\sin\overline{BAL}}{(AL)(AC)\sin\overline{LAC}} = \frac{AB\sin\overline{BAL}}{AC\sin\overline{LAC}}.$$

1.1.10 THEOREM. *If a, b, c, d are four distinct lines passing through a point V, then*

$$(\sin\overline{AVC}/\sin\overline{CVB})/(\sin\overline{AVD}/\sin\overline{DVB})$$

is independent of the positions of A, B, C, D on the lines a, b, c, d, respectively, so long as they are all distinct from V.

The value of the expression certainly will not change if any one of the points is taken at a different position of its line on the same side of V. The reader can easily show that the value of the expression also will not change if any one of the points is taken at a position of its line on the opposite side of V.

We now close the present section with the following convenient definitions.

1.1.11 Definitions. Straight lines that lie in a plane and pass through a common point are said to be *concurrent*. A set of concurrent coplanar lines is said to constitute a *pencil* of lines, and the point through which they all pass is called the *vertex* of the pencil. A pencil consisting of all the lines through its vertex is called a *complete pencil*. A line in the plane of a pencil and not passing through the vertex of the pencil is called a *transversal* of the pencil.

PROBLEMS

1. If $A_1, A_2, \ldots, A_n$ are n collinear points, show that

$$\overline{A_1A_2} + \overline{A_2A_3} + \cdots + \overline{A_{n-1}A_n} + \overline{A_nA_1} = 0.$$

2. If A, B, P are collinear and M is the midpoint of AB, show that $\overline{PM} = (\overline{PA} + \overline{PB})/2$.

3. If O, A, B are collinear, show that $\overline{OA}^2 + \overline{OB}^2 = \overline{AB}^2 + 2(\overline{OA})(\overline{OB})$.

4. If O, A, B, C are collinear and $\overline{OA} + \overline{OB} + \overline{OC} = 0$ and if P is any point on the line AB, show that $\overline{PA} + \overline{PB} + \overline{PC} = 3\,\overline{PO}$.

5. If on the same line we have $\overline{OA} + \overline{OB} + \overline{OC} = 0$ and $\overline{O'A'} + \overline{O'B'} + \overline{O'C'} = 0$, show that $\overline{AA'} + \overline{BB'} + \overline{CC'} = 3\,\overline{OO'}$.

6. If A, B, C are collinear and P, Q, R are the midpoints of BC, CA, AB respectively, show that the midpoints of CR and PQ coincide.

7. Let a and b be two given (positive) segments. Construct points P and Q on AB such that $\overline{AP}/\overline{PB} = a/b$ and $\overline{AQ}/\overline{QB} = -a/b$.

8. Show that if two points divide a line segment AB in equal ratios, then the two points coincide.

9. If r denotes the ratio $\overline{OA}/\overline{OB}$ and r' the ratio $\overline{OA'}/\overline{OB'}$, where O, A, B, A', B' are collinear, show that

$$rr'\,\overline{BB'} + r\overline{A'B} + r'\,\overline{B'A} + \overline{AA'} = 0.$$

10. If M is the midpoint of side BC of triangle ABC, and if $AB < AC$, prove that $\sphericalangle MAC < \sphericalangle BAM$.

11. If AL is the bisector of angle A in triangle ABC, show that $\overline{BL}/\overline{LC} = AB/AC$.

12. If AL is the bisector of exterior angle A of triangle ABC, where $AB \neq AC$, *show that* $\overline{BL}/\overline{LC} = -AB/AC$.

13. Prove *Stewart's Theorem*: If A, B, C are any three points on a line and P any point, then $\overline{PA}^2 \cdot \overline{BC} + \overline{PB}^2 \cdot \overline{CA} + \overline{PC}^2 \cdot \overline{AB} + \overline{BC} \cdot \overline{CA} \cdot \overline{AB} = 0$. (This theorem was stated, without proof, by Matthew Stewart (1717–1785) in 1746; it was rediscovered and proved by Thomas Simpson (1710–1761) in 1751, by L. Euler in 1780, and by L. N. M. Carnot in 1803. The case where P lies on the line ABC is found in Pappus' *Mathematical Collection*, ca. 300.)

14. Find the lengths of the medians of a triangle having sides a, b, c.

15. Find the lengths of the angle bisectors of a triangle having sides a, b, c.

16. Give a direct proof of the *Steiner-Lehmus Theorem*: If the bisectors of the base angles of a triangle are equal, the triangle is isosceles. (This problem was proposed in 1840 by D. C. Lehmus (1780–1863) to Jacob Steiner (1796–1863).)

17. Show that the sum of the squares of the distances of the vertex of the right angle of a right triangle from the two points of trisection of the hypotenuse is equal to 5/9 the square of the hypotenuse.

18. If A, B, C are three collinear points and a, b, c are the tangents from A, B, C to a given circle, then

$$a^2\overline{BC} + b^2\overline{CA} + c^2\overline{AB} + \overline{BC} \cdot \overline{CA} \cdot \overline{AB} = 0.$$

19. If A, B, C, D, O are any five coplanar points, show that
(a) $\sin \overline{AOD} \sin \overline{BOC} + \sin \overline{BOD} \sin \overline{COA} + \sin \overline{COD} \sin \overline{AOB} = 0$.
(b) $\Delta\overline{AOD}\, \Delta\overline{BOC} + \Delta\overline{BOD}\, \Delta\overline{COA} + \Delta\overline{COD}\, \Delta\overline{AOB} = 0$.

20. Using 19(a), prove *Ptolemy's Theorem*: If $ABCD$ is a cyclic quadrilateral, then $AD \cdot BC + AB \cdot CD = AC \cdot BD$.

21. If O is any point in the plane of triangle ABC, show that

$$\Delta\overline{OBC} + \Delta\overline{OCA} + \Delta\overline{OAB} = \Delta\overline{ABC}.$$

22. If P is any point in the plane of the parallelogram $ABCD$, show that $\Delta\overline{PAB} + \Delta\overline{PCD} = \Delta\overline{ABC}$.

23. If A, B, C, D, P, Q are any six distinct collinear points, show that

$$(\overline{AP} \cdot \overline{AQ})/(\overline{AB} \cdot \overline{AC} \cdot \overline{AD}) + (\overline{BP} \cdot \overline{BQ})/(\overline{BC} \cdot \overline{BD} \cdot \overline{BA}) + (\overline{CP} \cdot \overline{CQ})/(\overline{CD} \cdot \overline{CA} \cdot \overline{CB}) + (\overline{DP} \cdot \overline{DQ})/(\overline{DA} \cdot \overline{DB} \cdot \overline{DC}) = 0.$$

24. Generalize Problem 23 for n points $A, B, \ldots$ and $n - 2$ points $P, Q, \ldots$.

25. Complete the proof of Theorem 1.1.10.

1.2 INFINITE ELEMENTS

Another innovation of modern elementary geometry is the creation of some ideal elements called "points at infinity," "the line at infinity" in a plane, and "the plane at infinity" in space. The purpose of introducing these ideal elements is to eliminate certain bothersome case distinctions in plane and solid geometry which arise from the possibility of lines and planes being either parallel or intersecting. It follows that with these ideal elements many theorems can be given a single universal statement, whereas without these ideal elements the statements have to be qualified to take care of various exceptional situations. Subterfuges of this sort are common in mathematics. Consider, for example, the discussion of quadratic equations in elementary algebra. The equation $x^2 - 2x + 1 = 0$ actually has only the one root, $x = 1$, but for the sake of uniformity it is agreed to say that the equation has *two equal roots*, each equal to 1. Again, in order that the equation $x^2 + x + 1 = 0$ have any root at all, it is agreed to extend the number system so as to include imaginary numbers. With these two conventions—that a repeated root is to count as two roots and that imaginary roots are to be accepted equally with real roots—we can assert as a universal statement that "every quadratic equation with real coefficients has exactly two roots."

The introduction into geometry of the notion of points at infinity is usually credited to Johann Kepler (1571–1630), but it was Gérard Desargues (1593–1662) who, in a treatment of the conic sections (his *Brouillon projet*) published in 1639, first used the idea systematically. This work of Desargues marks the first essential advance in synthetic geometry since the time of the ancient Greeks.

Restricting ourselves for the time being to plane geometry, consider two lines l_1 and l_2, where l_1 is held fast while l_2 rotates about a fixed point O in the plane but not on line l_1. As l_2 approaches the position of parallelism with l_1, the point P of intersection of l_1 and l_2 recedes farther and farther along line l_1, and in the limiting position of parallelism the point P ceases to exist. To accommodate this exceptional situation, we agree to augment the set of points on l_1 by an ideal point, called *the point at infinity* on l_1, and we say that when l_1 and l_2 are parallel they intersect in this ideal point on l_1.

If our introduction of an ideal point at infinity on a line is not to create more exceptions than it removes, it must be done in such a way that two distinct points, ordinary or ideal, determine one and only one line, and such that two distinct lines intersect in one and only one point. As a first consequence of this we see that two parallel lines must intersect in the same ideal point, no matter in which direction the lines are traversed.

Suppose l_1 and l_2 are two parallel lines intersecting in the ideal point I, and let O be any ordinary point not on l_1 or l_2. Since O and I are to determine a line l_3, and since l_3 cannot intersect l_1 or l_2 a second time, we see that l_3 must be the parallel to l_1 and l_2 through point O. That is, the ideal point I lies on all three of the parallel lines l_1, l_2, l_3, and, by the same argument, on all lines parallel to l_1. Thus the members of a family of parallel lines must share a common ideal point at infinity.

It is easy to see that a different ideal point must be assigned for a different family of parallel lines. Let line m_1 cut line l_1 in an ordinary point P, and let m_2 be a line parallel to m_1. Then m_1 and m_2 intersect in an ideal point J which must be distinct from I, since otherwise the distinct lines m_1 and l_1 would intersect in the two points P and I.

Now consider two distinct ideal points, I and J. The line l which they are to determine cannot pass through any ordinary point P of the plane. For if it did, then line l_1 determined by P and I and line l_2 determined by P and J would be a pair of distinct ordinary lines, each of which would be contained in line l because of the collinearity of P, I, J. It follows that the line l determined by a pair of ideal points can contain only ideal points and must therefore be an ideal line, which we call a *line at infinity*.

Finally, we see that in the plane there can be only one line at infinity. For if l_1 and l_2 should be two distinct lines at infinity, these lines would have to intersect in an ideal point I. A line l_3 passing through an ordinary point O and not passing through I would have to intersect l_1 and l_2 in distinct ideal points J and K respectively. The line through J and K would then contain the ordinary point O, which we have seen is impossible.

The above discussion leads to the following convention and theorem.

1.2.1 CONVENTION AND DEFINITIONS. We agree to add to the points of the plane a collection of ideal points, called *points at infinity*, such that

1. each ordinary line of the plane contains exactly one ideal point,
2. the members of a family of parallel lines in the plane share a common ideal point, distinct families having distinct ideal points.

The collection of added ideal points is regarded as an ideal line, called the *line at infinity*, which contains no ordinary points.

The plane, augmented by the above ideal points, will be referred to as the *extended plane*.

1.2.2 THEOREM. *In the extended plane, any two distinct points determine one and only one line and any two distinct lines intersect in one and only one point.*

If a point P recedes indefinitely along the line determined by two ordinary points A and B, then $\overline{AP}/\overline{PB}$ approaches the limiting value -1. This motivates the following definition.

1.2.3 DEFINITION. If A and B are any two ordinary points, and I the ideal point, on a given line, then we define $\overline{AI}/\overline{IB}$ to be -1.

The reader may care to supply an analysis leading to the following convention, definitions, and theorem for three-dimensional space.

1.2.4 CONVENTION AND DEFINITIONS. We agree to add to the points of space a collection of ideal points, called *points at infinity*, and ideal lines, called *lines at infinity*, such that

1. each ordinary line of space contains exactly one ideal point,
2. the members of a family of parallel lines in space share a common ideal point, distinct families having distinct ideal points,
3. each ordinary plane of space contains exactly one ideal line,
4. the members of a family of parallel planes in space share a common ideal line, distinct families having distinct ideal lines,
5. the ideal line of an ordinary plane in space consists of the ideal points of the ordinary lines of that plane.

The collection of added ideal points and ideal lines is regarded as an ideal plane, called the *plane at infinity*, which contains no ordinary points or lines.

Three-dimensional space augmented by the above ideal elements, will be referred to as *extended three-dimensional* space.

1.2.5 THEOREM. *In extended three-dimensional space, any two distinct coplanar lines intersect in one and only one point, any non-incident line and plane intersect in one and only one point, any three non-coaxial planes (that is, planes not sharing a common line) intersect in one and only one point, any two disinct planes intersect in one and only one line, any two distinct points determine one and only one line, any two distinct intersecting lines determine one and only one plane, any non-incident point and line determine one and only one plane, any three noncollinear points determine one and only one plane.*

There still remains the important matter of whether the above conventions can ever lead to a contradiction. One is reminded of efforts to extend the concept of quotient so as to include fractions having zero denominators. It can be shown — but we lack the space to do so — that our conventions about infinite elements cannot lead to any contradiction.

PROBLEMS

1. Develop an analysis leading to Convention 1.2.4 and Theorem 1.2.5.

2. Which of the following statements are true for the ordinary plane and which are true for the extended plane?
   (a) There is a one-to-one correspondence between the lines through a fixed

point O and the points of a fixed line l, not passing through O, such that corresponding lines and points are in incidence.
(b) The bisector of an exterior angle of an ordinary triangle divides the opposite side externally in the ratio of the adjacent sides.
(c) Every straight line possesses one and only one ideal point.
(d) Every straight line possesses infinitely many ordinary points.
(e) If a triangle is the figure determined by any three non-concurrent straight lines, then every triangle encloses a finite area.
(f) If parallel lines are lines lying in the same plane and having no ordinary point in common, then through a given point O there passes one and only one line parallel to a given line l not containing O.
(g) If A and B are distinct ordinary points and r is any real number, there is a unique point P on line AB such that $\overline{AP}/\overline{PB} = r$.

3. Translate the following theorems of ordinary three-dimensional space into the language of infinite elements, and then supply simple proofs.
(a) Through a given point there is one and only one plane parallel to a given plane not containing the given point.
(b) Two (distinct) lines which are parallel to a third line are parallel to each other.
(c) If a line is parallel to each of two intersecting planes, it is parallel to their line of intersection.
(d) If a line is parallel to the line of intersection of two intersecting planes, then it is parallel to each of the two planes.
(e) If a line l is parallel to a plane p, then any plane containing l and intersecting p, cuts p in a line parallel to l.
(f) Through a given line one and only one plane can be passed parallel to a given skew line.
(g) Through a given point one and only one plane can be passed parallel to each of two skew lines, neither of which contains the given point.
(h) All the lines through a point and parallel to a given plane lie in a plane parallel to the first plane.
(i) If a plane contains one of two parallel lines but not the other, then it is parallel to the other.
(j) The intersections of a plane with two parallel planes are parallel lines.

4. Define prism and cylinder in terms of pyramid and cone respectively.

5. If A, B, C, P, Q are any five collinear points, show that

$$(\overline{AP} \cdot \overline{AQ})/(\overline{AB} \cdot \overline{AC}) + (\overline{BP} \cdot \overline{BQ})/(\overline{BC} \cdot \overline{BA}) + (\overline{CP} \cdot \overline{CQ})/(\overline{CA} \cdot \overline{CB}) = 1.$$

6. If A, B, C, D, P are any five collinear points, show that

$$\overline{AP}/(\overline{AB} \cdot \overline{AC} \cdot \overline{AD}) + \overline{BP}/(\overline{BC} \cdot \overline{BD} \cdot \overline{BA}) + \overline{CP}/(\overline{CD} \cdot \overline{CA} \cdot \overline{CB}) + \overline{DP}/(\overline{DA} \cdot \overline{DB} \cdot \overline{DC}) = 0.$$

7. Generalize Problem 5 for n points $A, B, \ldots$ and $n-1$ points $P, Q, \ldots$.

8. Generalize Problem 6 for n points A, B, . . . and $n-k$ points P, Q, . . . , where $1 < k < n-1$.

1.3 THE THEOREMS OF MENELAUS AND CEVA

The theorems of Menelaus and Ceva, in their original versions, are quite old, for the one dates back to ancient Greece and the other to 1678. It is when they are stated in terms of sensed magnitudes that they assume a particularly modern appearance.

Menelaus of Alexandria was a Greek astronomer who lived in the first century A.D. Though his works in their original Greek are all lost to us, we know of some of them from remarks made by later commentators, and his three-book treatise *Sphaerica* has been preserved for us in the Arabic. This work throws considerable light on the Greek development of trigonometry. Book I is devoted to establishing for spherical triangles many of the propositions of Euclid's *Elements* that hold for plane triangles, such as the familiar congruence theorems, theorems about isosceles triangles, and so on. In addition, Menelaus establishes the congruence of two spherical triangles having the angles of one equal to the angles of the other (for which there is no analogue in the plane) and the fact that the sum of the angles of a spherical triangle is greater than two right angles. Book II contains theorems of interest in astronomy. Book III develops the spherical trigonometry of the time, largely deduced from the spherical analogue of the plane proposition now commonly referred to as *Menelaus' Theorem*. Actually, the plane case is assumed by Menelaus as well known and is used by him to establish the spherical case. A good deal of spherical trigonometry can be deduced from the spherical version of the theorem by taking special triangles and special transversals. L. N. M. Carnot made the theorem of Menelaus basic in his *Essai sur la théorie des transversales* of 1806.

Though the theorem of Ceva is a close companion theorem to that of Menelaus, it seems to have eluded discovery until 1678, when the Italian Giovanni Ceva (ca. 1647–1736) published a work containing both it and the then apparently long forgotten theorem of Menelaus.

The theorems of Menelaus and Ceva, in their modern dress, are powerful theorems, and they deal elegantly with many problems involving collinearity of points and concurrency of lines. We now turn to a study of these two remarkable theorems. The reader should observe how the convention as to points at infinity eliminates the separate consideration of a number of otherwise exceptional situations.

1.3.1 DEFINITIONS. We call a triangle *ordinary* if its three sides are in the unextended plane. An ordinary or ideal point lying on a side line of an ordinary triangle, but not coinciding with a vertex of the triangle, will be called a *menelaus point* of the triangle for this side.

1.3.2 MENELAUS' THEOREM. *A necessary and sufficient condition for three menelaus points D, E, F for the sides BC, CA, AB of an ordinary* *triangle ABC to be collinear is that*

$$(\overline{BD}/\overline{DC})(\overline{CE}/\overline{EA})(\overline{AF}/\overline{FB}) = -1.$$

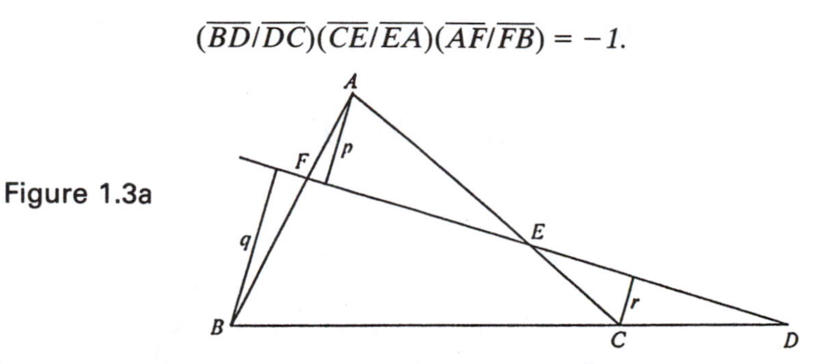

Figure 1.3a

Necessity. Suppose (see Figure 1.3a) D, E, F are collinear on a line l which is not the line at infinity. Drop perpendiculars p, q, r on l from A, B, C. Then, disregarding signs,

$$BD/DC = q/r, \quad CE/EA = r/p, \quad AF/FB = p/q.$$

It follows that

$$(\overline{BD}/\overline{DC})(\overline{CE}/\overline{EA})(\overline{AF}/\overline{FB}) = \pm 1.$$

Since, however, l must cut one or all three sides externally, we see that we can have only the – sign. If l is the line at infinity, the proof is simple.

Sufficiency. Suppose

$$(\overline{BD}/\overline{DC})(\overline{CE}/\overline{EA})(\overline{AF}/\overline{FB}) = -1$$

and let EF cut BC in D'. Then D' is a menelaus point and, by the above,

$$(\overline{BD'}/\overline{D'C})(\overline{CE}/\overline{EA})(\overline{AF}/\overline{FB}) = -1.$$

It follows that $\overline{BD}/\overline{DC} = \overline{BD'}/\overline{D'C}$, or that $D \equiv D'$. That is, D, E, F are collinear.

1.3.3 TRIGONOMETRIC FORM OF MENELAUS' THEOREM. *A necessary and sufficient condition for three menelaus points D, E, F for the sides BC, CA, AB of an ordinary triangle ABC to be collinear is that*

$$(\sin \overline{BAD}/\sin \overline{DAC})(\sin \overline{CBE}/\sin \overline{EBA})(\sin \overline{ACF}/\sin \overline{FCB}) = -1.$$

For we have, by Theorem 1.1.9,

$$\overline{BD}/\overline{DC} = (AB \sin \overline{BAD})/(AC \sin \overline{DAC}),$$
$$\overline{CE}/\overline{EA} = (BC \sin \overline{CBE})/(BA \sin \overline{EBA}),$$
$$\overline{AF}/\overline{FB} = (CA \sin \overline{ACF})/(CB \sin \overline{FCB}).$$

It follows that

$$(\sin \overline{BAD}/\sin \overline{DAC})(\sin \overline{CBE}/\sin \overline{EBA})(\sin \overline{ACF}/\sin \overline{FCB}) = -1$$

if and only if

$$(\overline{BD}/\overline{DC})(\overline{CE}/\overline{EA})(\overline{AF}/\overline{FB}) = -1.$$

Hence the theorem.

1.3.4 DEFINITION. A line passing through a vertex of an ordinary triangle, but not coinciding with a side of the triangle, will be called a *cevian line* of the triangle for this vertex. A cevian line will be identified by the vertex to which it belongs and the point in which it cuts the opposite side, as cevian line AD through vertex A of triangle ABC and cutting the opposite side BC in the point D.

1.3.5 CEVA'S THEOREM. *A necessary and sufficient condition for three cevian lines AD, BE, CF of an ordinary triangle ABC to be concurrent is that*

$$(\overline{BD}/\overline{DC})(\overline{CE}/\overline{EA})(\overline{AF}/\overline{FB}) = +1.$$

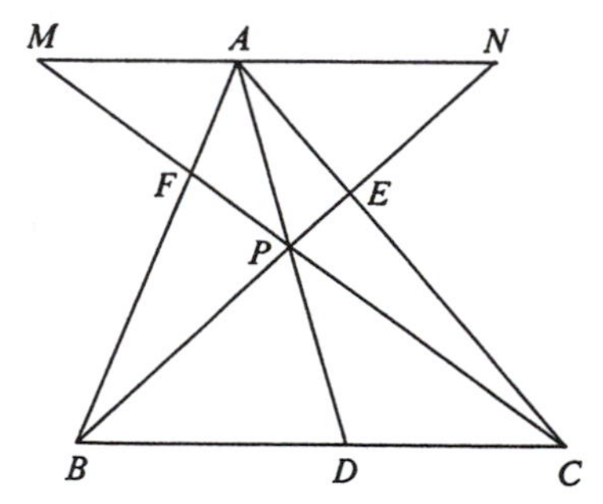

Figure 1.3b

Necessity. Suppose (see Figure 1.3b) AD, BE, CF are concurrent in P. Without loss of generality we may assume that P does not lie on the parallel through A to BC. Let BE, CF intersect this parallel in N and M. Then, disregarding signs,

$$BD/DC = AN/MA, \quad CE/EA = BC/AN, \quad AF/FB = MA/BC,$$

whence

$$(\overline{BD}/\overline{DC})(\overline{CE}/\overline{EA})(\overline{AF}/\overline{FB}) = \pm 1.$$

That the sign must be + follows from the fact that either none or two of the points D, E, F divide their corresponding sides externally.

Sufficiency. Suppose

$$(\overline{BD}/\overline{DC})(\overline{CE}/\overline{EA})(\overline{AF}/\overline{FB}) = +1$$

and let BE, CF intersect in P and draw AP to cut BC in D'. Then AD' is a cevian line. Hence, by the above, we have

$$(\overline{BD}'/\overline{D'C})(\overline{CE}/\overline{EA})(\overline{AF}/\overline{FB}) = +1.$$

It follows that $\overline{BD}'/\overline{D'C} = \overline{BD}/\overline{DC}$, or that $D \equiv D'$. That is, AD, BE, CF are concurrent.

1.3.6 TRIGONOMETRIC FORM OF CEVA'S THEOREM. *A necessary and sufficient condition for three cevian lines AD, BE, CF of an ordinary triangle ABC to be concurrent is that*

$$(\sin \overline{BAD}/\sin \overline{DAC})(\sin \overline{CBE}/\sin \overline{EBA})(\sin \overline{ACF}/\sin \overline{FCB}) = +1.$$

The reader can easily supply a proof similar to that given for Theorem 1.3.3.

PROBLEMS

1. Supply a proof of Theorem 1.3.6.

2. Prove the "necessary" part of Menelaus' Theorem by drawing a line (see Figure 1.3a) through C parallel to DEF to cut AB in L.

3. Prove the "necessary" part of Menelaus' Theorem using the feet of the perpendiculars from the vertices of the triangle on any line perpendicular to the transversal.

4. Derive the "necessary" part of Ceva's Theorem by applying Menelaus' Theorem (see Figure 1.3b) to triangle ABD with transversal CPF and to triangle ADC with transversal BPE.

5. Points E and F are taken on the sides CA, AB of a triangle ABC such that $\overline{CE}/\overline{EA} = \overline{AF}/\overline{FB} = k$. If EF cuts BC in D, show that $\overline{CD} = k^2\overline{BD}$.

6. In Figure 1.3b, prove that $PD/AD + PE/BE + PF/CF = 1$.

7. In Figure 1.3b, prove that $AP/AD + BP/BE + CP/CF = 2$.

8. In Figure 1.3b, prove that $\overline{AF}/\overline{FB} + \overline{AE}/\overline{EC} = \overline{AP}/\overline{PD}$.

9. If the sides AB, BC, CD, DA of a quadrilateral $ABCD$ are cut by a transversal in the points A', B', C', D' respectively, show that

$$(\overline{AA}'/\overline{A'B})(\overline{BB}'/\overline{B'C})(\overline{CC}'/\overline{C'D})(\overline{DD}'/\overline{D'A}) = 1.$$

10. If a transversal cuts the sides AB, BC, CD, DE, ... of an n-gon $ABCDE$ ... in the points A', B', C', D', ..., show that

$$(\overline{AA}'/\overline{A'B})(\overline{BB}'/\overline{B'C})(\overline{CC}'/\overline{C'D})(\overline{DD}'/\overline{D'E}) \cdots = (-1)^n.$$

(This is a generalization of Menelaus' Theorem.)

11. If on the sides AB, BC, CD, DA of a quadrilateral $ABCD$, points A', B', C', D' are taken such that

$$\overline{AA'} \cdot \overline{BB'} \cdot \overline{CC'} \cdot \overline{DD'} = \overline{A'B} \cdot \overline{B'C} \cdot \overline{C'D} \cdot \overline{D'A},$$

show that $A'B'$ and $C'D'$ intersect on AC, and $A'D'$ and $B'C'$ intersect on BD.

12. Let the sides AB, BC, CD, DA of a nonplanar quadrilateral $ABCD$ be cut by a plane in the points A', B', C', D'. Show that

$$\overline{AA'} \cdot \overline{BB'} \cdot \overline{CC'} \cdot \overline{DD'} = \overline{A'B} \cdot \overline{B'C} \cdot \overline{C'D} \cdot \overline{D'A}.$$

13. Let D', E', F' be menelaus points on the sides $B'C'$, $C'A'$, $A'B'$ of a triangle $A'B'C'$, and let O be a point in space not in the plane of triangle $A'B'C'$. Show that points D', E', F' are collinear if and only if

$$(\sin \overline{B'OD'} / \sin \overline{D'OC'})(\sin \overline{C'OE'} / \sin \overline{E'OA'})(\sin \overline{A'OF'} / \sin \overline{F'OB'}) = -1.$$

14. Let D, E, F be three menelaus points on the sides BC, CA, AB of a spherical triangle ABC. Show that D, E, F lie on a great circle of the sphere if and only if

$$(\sin \overset{\frown}{BD} / \sin \overset{\frown}{DC})(\sin \overset{\frown}{CE} / \sin \overset{\frown}{EA})(\sin \overset{\frown}{AF} / \sin \overset{\frown}{FB}) = -1.$$

(This is the theorem that Menelaus used in Book III of his *Sphaerica*.)

15. If the lines joining a point O to the vertices of a polygon $ABCD \ldots$ of an odd number of sides meet the opposite sides AB, BC, CD, $DE, \ldots$ in the points A', B', C', $D', \ldots$, show that

$$(\overline{AA'} / \overline{A'B})(\overline{BB'} / \overline{B'C})(\overline{CC'} / \overline{C'D})(\overline{DD'} / \overline{D'E}) \cdots = 1.$$

(This is a generalization of Ceva's Theorem.)

1.4 APPLICATIONS OF THE THEOREMS OF MENELAUS AND CEVA

We illustrate the power of the theorems of Menelaus and Ceva by now using them to establish three useful and highly attractive theorems. Many further illustrations will be found among the problems at the end of this section.

1.4.1 THEOREM. *If AD, BE, CF are any three concurrent cevian lines of an ordinary triangle ABC, and if D' denotes the point of intersection of BC and FE, then D and D' divide BC, one internally and one externally, in the same numerical ratio.*

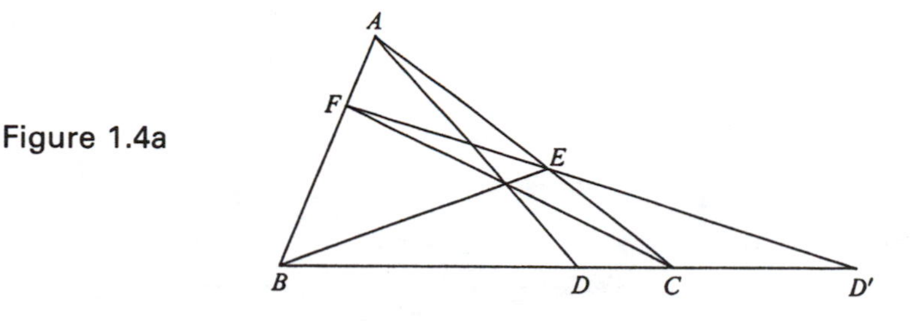

Figure 1.4a

Since AD, BE, CF are concurrent cevian lines (see Figure 1.4a) we have, by Ceva's Theorem,

$$(\overline{BD}/\overline{DC})(\overline{CE}/\overline{EA})(\overline{AF}/\overline{FB}) = +1.$$

Since D', E, F are collinear menelaus points we have, by Menelaus' Theorem,

$$(\overline{BD'}/\overline{D'C})(\overline{CE}/\overline{EA})(\overline{AF}/\overline{FB}) = -1.$$

It follows that

$$\overline{BD}/\overline{DC} = -\overline{BD'}/\overline{D'C},$$

whence D and D' divide BC, one internally and one externally, in the same numerical ratio.

1.4.2 Definitions. Two triangles ABC and $A'B'C'$ are said to be *copolar* if AA', BB', CC' are concurrent; they are said to be *coaxial* if the points of intersection of BC and $B'C'$, CA and $C'A'$, AB and $A'B'$ are collinear.

1.4.3 Desargues' two-triangle theorem. *Copolar triangles are coaxial, and conversely.*

Let the two triangles (see Figure 1.4b) be ABC and $A'B'C'$. Suppose AA', BB', CC' are concurrent in a point O. Let P, Q, R be the points of intersection of BC and $B'C'$, CA and $C'A'$, AB and $A'B'$. Considering the triangles BCO, CAO, ABO in turn, with the respective transversals $B'C'P$, $C'A'Q$, $A'B'R$, we find, by Menelaus' Theorem,

$$(\overline{BP}/\overline{PC})(\overline{CC'}/\overline{C'O})(\overline{OB'}/\overline{B'B}) = -1,$$
$$(\overline{CQ}/\overline{QA})(\overline{AA'}/\overline{A'O})(\overline{OC'}/\overline{C'C}) = -1,$$
$$(\overline{AR}/\overline{RB})(\overline{BB'}/\overline{B'O})(\overline{OA'}/\overline{A'A}) = -1.$$

Setting the product of the three left members of the above equations equal to the product of the three right members, we obtain

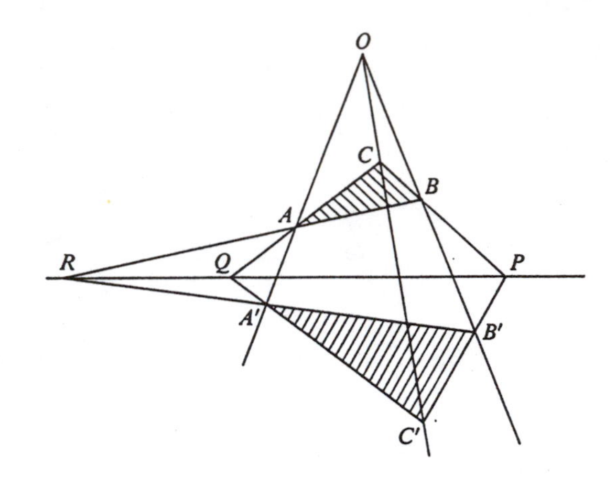

Figure 1.4b

$$(\overline{BP}/\overline{PC})(\overline{CQ}/\overline{QA})(\overline{AR}/\overline{RB}) = -1,$$

whence P, Q, R are collinear. Thus copolar triangles are coaxial.

Conversely, suppose P, Q, R are collinear and let O be the point of intersection of AA' and BB'. Now triangles AQA' and BPB' are copolar, and therefore coaxial. That is, O, C, C' are collinear. Thus coaxial triangles are copolar.

1.4.4 Pascal's "mystic hexagram" theorem for a circle. *The points L, M, N of intersection of the three pairs of opposite sides AB and DE, BC and EF, FA and CD of a (not necessarily convex) hexagon $ABCDEF$ inscribed in a circle lie on a line, called the Pascal line of the hexagon.*

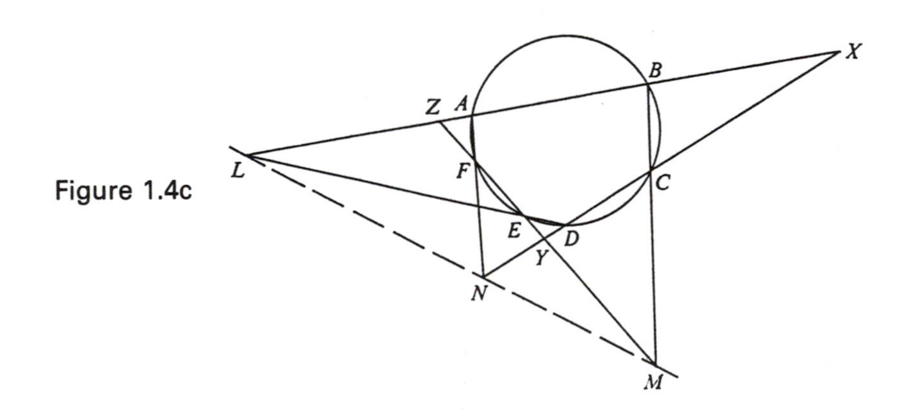

Figure 1.4c

Let X, Y, Z (see Figure 1.4c) be the points of intersection of AB and CD, CD and EF, EF and AB, and consider DE, FA, BC as transversals cutting the sides of triangle XYZ. By the Theorem of Menelaus we have

$$(\overline{XL}/\overline{LZ})(\overline{ZE}/\overline{EY})(\overline{YD}/\overline{DX}) = -1,$$
$$(\overline{XA}/\overline{AZ})(\overline{ZF}/\overline{FY})(\overline{YN}/\overline{NX}) = -1,$$
$$(\overline{XB}/\overline{BZ})(\overline{ZM}/\overline{MY})(\overline{YC}/\overline{CX}) = -1.$$

Setting the product of the three left members of the above equations equal to the product of the three right members and rearranging the ratios, we obtain

$$(1) \qquad \left(\frac{\overline{XL}}{\overline{LZ}} \cdot \frac{\overline{ZM}}{\overline{MY}} \cdot \frac{\overline{YN}}{\overline{NX}}\right)\left(\frac{\overline{XB} \cdot \overline{XA}}{\overline{XC} \cdot \overline{XD}}\right)\left(\frac{\overline{YC} \cdot \overline{YD}}{\overline{YE} \cdot \overline{YF}}\right)\left(\frac{\overline{ZE} \cdot \overline{ZF}}{\overline{ZB} \cdot \overline{ZA}}\right) = -1.$$

But

$$\overline{XB} \cdot \overline{XA} = \overline{XC} \cdot \overline{XD},$$
$$\overline{YC} \cdot \overline{YD} = \overline{YE} \cdot \overline{YF},$$
$$\overline{ZE} \cdot \overline{ZF} = \overline{ZB} \cdot \overline{ZA},$$

whence each of the last three factors in parentheses in (1) has the value 1. It follows that

$$(\overline{XL}/\overline{LZ})(\overline{ZM}/\overline{MY})(\overline{YN}/\overline{NX}) = -1,$$

or L, M, N are collinear.

Should vertex X, say, of triangle XYZ be an ideal point, so that CD is parallel to AB, the above proof fails. But in this case we may choose a point C' on the circle and near to C, and consider the hexagon $ABC'DEF$ as C' approaches C along the circle. Since the Pascal line exists for all positions of C' as it approaches C, the Pascal line also exists in the limit.

Desargues' two-triangle theorem appears to have been given by Desargues in a work on perspective in 1636, three years before his *Brouillon projet* was published. This theorem has become basic in the present-day theory of projective geometry, and we shall meet it again in later chapters. In deeper work one encounters so-called *non-Desarguesian geometries*, or plane geometries in which the two-triangle theorem fails to hold. The great French geometer Jean-Victor Poncelet (1788–1867) made Desargues' two-triangle theorem the foundation of his theory of homologic figures.

Blaise Pascal (1623–1662) was inspired by the work of Desargues and was in possession of his "mystic hexagram" theorem for a general conic when he was only 16 years old. The consequences of the "mystic hexagram" theorem are very numerous and attractive, and an almost unbelievable amount of research has been expended on the configuration. There are $5!/2 = 60$ possible ways of forming a hexagon from 6 points on a circle, and, by Pascal's theorem, to each hexagon corresponds a *Pascal line*. These 60 Pascal lines pass three by three through 20 points, called *Steiner points*, which in turn lie four by four on 15 lines, called *Plücker lines*. The Pascal lines also concur three by three in another set of points, called *Kirkman points*, of which there are 60. Corresponding to each Steiner point, there are three Kirkman points such that all four lie upon a line, called a *Cayley line*. There are 20 of these Cayley lines, and they pass four by four through 15 points, called *Salmon points*. There are many further extensions and properties of the configuration, and the number of different proofs that have been supplied for the "mystic hexagram" theorem itself is now legion. Some of these alternative proofs will be met in later parts of the book, as will some of the numerous corollaries of the theorem.

PROBLEMS

1. Using Ceva's Theorem prove that (a) The medians of a triangle are concurrent. (b) The internal angle bisectors of a triangle are concurrent. (c) The altitudes of a triangle are concurrent.

2. If D, E, F are the points of contact of the inscribed circle of triangle ABC with the sides BC, CA, AB respectively, show that AD, BE, CF are concurrent. (This point of concurrency is called the *Gergonne point* of the triangle, after J. D. Gergonne (1771–1859), founder-editor of the mathematics journal *Annales de mathématiques*. Just why the point was named after Gergonne seems not to be known.)

3. Let D, E, F be the points on the sides BC, CA, AB of triangle ABC such that D is halfway around the perimeter from A, E halfway around from B, and F halfway around from C. Show that AD, BE, CF are concurrent. (This point of concurrency is called the *Nagel point* of the triangle, after C. H. Nagel (1803–1882), who considered it in a work of 1836.)

4. Let X and X' be points on a line segment MN symmetric with respect to the midpoint of MN. Then X and X' are called a pair of *isotomic points* for the segment MN. Show that if D and D', E and E', F and F' are isotomic points for the sides BC, CA, AB of triangle ABC, and if AD, BE, CF are concurrent, then AD', BE', CF' are also concurrent. (Two such related points of concurrency are called a pair of *isotomic conjugate points* for the triangle, a term introduced by John Casey in 1889.)

5. Show that the Gergonne and Nagel points of a triangle are a pair of isotomic conjugate points for the triangle. (See Problems 2, 3, 4 for the required definitions.)

6. If, in Problem 4, D, E, F are collinear, show that D', E', F' are also collinear. (Two such related lines as DEF and $D'E'F'$ are sometimes called a pair of *reciprocal transversals* of the triangle ABC, a name used by G. de Longchamps in 1890.)

7. Let OX and OX' be rays through vertex O of angle MON symmetric with respect to the bisector of angle MON. Then OX and OX' are called a pair of *isogonal lines* for the angle MON. Show that if AD and AD', BE and BE',

CF and CF' are isogonal cevian lines for the angles A, B, C of a triangle ABC, and if AD, BE, CF are concurrent, then AD', BE', CF' are also concurrent. (This theorem was given by J. J. A. Mathieu in 1865, and was extended to three-space by J. Neuberg in 1884. Two such related points of concurrency are called a pair of *isogonal conjugate points* for the triangle. The orthocenter and circumcenter of a triangle are a pair of isogonal conjugate points. The incenter is its own isogonal conjugate. The isogonal conjugate of the centroid is called the *symmedian point* of the triangle; it enjoys some very attractive properties.)

8. If, in Problem 7, D, E, F are collinear, show that D', E', F' are also collinear.

9. Let AD, BE, CF be three concurrent cevian lines of triangle ABC, and let the circle through D, E, F intersect the sides BC, CA, AB again in D', E', F'. Show that AD', BE', CF' are concurrent.

10. Show that the tangents to the circumcircle of a triangle at the vertices of the triangle intersect the opposite sides of the triangle in three collinear points.

11. If AD, BE, CF are three cevian lines of an ordinary triangle ABC, concurrent in a point P, and if EF, FD, DE intersect the sides BC, CA, AB of triangle ABC in the points D', E', F', show that D', E', F' are collinear on a line p. (J. J. A. Mathieu, in 1865, called the point P the *trilinear pole* of the line p, and the line p the *trilinear polar* of the point P, for the triangle ABC. The trilinear polar of the orthocenter of a triangle is called the *orthic axis* of the triangle.)

12. Prove that the external bisectors of the angles of a triangle intersect the opposite sides in three collinear points.

13. Prove that two internal angle bisectors and the external bisector of the third angle of a triangle intersect the opposite sides in three collinear points.

14. Two parallelograms $ACBD$ and $A'CB'D'$ have a common angle at C. Prove that DD', $A'B$, AB' are concurrent.

15. Let $ABCD$ be a parallelogram and P any point. Through P draw lines parallel to BC and AB to cut BA and CD in G and H and AD and BC in E and F. Prove that the diagonal lines EG, HF, DB are concurrent.

16. If equilateral triangles BCA', CAB', ABC' are described externally upon the sides BC, CA, AB of triangle ABC, show that AA', BB', CC' are concurrent in a point P. (The point P is the first notable point of the triangle discovered after Greek times. If the angles of triangle ABC are each less than 120°, then P is the point the sum of whose distances from A, B, C is a minimum. The minimization problem was proposed to Torricelli by Fermat. Torricelli solved the problem and his solution was published in 1659 by his student Viviani.)

17. Show that, in Problem 16, AA', BB', CC' are still concurrent if the equilateral triangles are described internally upon the sides of the given triangle ABC. (The two points of concurrency of Problems 16 and 17 are known as the *isogonic centers* of triangle ABC. The isogonal conjugates of the isogonic

centers are called the *isodynamic points* of the triangle, a term given by J. Neuberg in 1885.)

18. Let ABC be a triangle right-angled at B, and let $BCDD'$ and $BAEE'$ be squares drawn on BC and BA externally to the triangle ABC. Prove that CE and AD intersect on the altitude of triangle ABC through B.

19. If the sides of a triangle ABC are cut by a transversal in D, E, F, all exterior to the circumcircle of the triangle, show that the product of the tangent lengths from D, E, F to the circumcircle is equal to $AF \cdot BD \cdot CE$.

20. A transversal cuts the sides BC, CA, AB of a triangle ABC in D, E, F. P, Q, R are the midpoints of EF, FD, DE, and AP, BQ, CR intersect BC, CA, AB in X, Y, Z. Show that X, Y, Z are collinear.

21. Let O and U be two points in the plane of triangle ABC. Let AO, BO, CO intersect the opposite sides BC, CA, AB in P, Q, R. Let PU, QU, RU intersect QR, RP, PQ respectively in X, Y, Z. Show that AX, BY, CZ are concurrent.

22. Let AA', BB', CC' be three cevian lines for triangle ABC. Let 1 be a point on BC, 2 the intersection of $1B'$ and BA, 3 of $2A'$ and AC, 4 of $3C'$ and CB, 5 of $4B'$ and BA, 6 of $5A'$ and AC, 7 of $6C'$ and CB. Show that point 7 coincides with point 1. (This interesting closure theorem is due to O. Nehring, 1942.)

1.5 CROSS RATIO

Another topic which originated with the ancient Greeks, and of which certain aspects were very fully investigated by them, is that now called "cross ratio." In the nineteenth century this concept was revived and considerably improved with the aid of sensed magnitudes and with a highly convenient notation rendered possible by sensed magnitudes. The modern development of the subject is due, independently of each other, to Möbius (in his *Der barycentrische Calcul* of 1827) and Michel Chasles (in his *Aperçu historique sur l'origine et le développement des méthodes en géométrie* of 1829–1837, his *Traité de géométrie supérieure* of 1852, and his *Traité des sections coniques* of 1865). A treatment of the cross-ratio concept freed of metrical considerations was made by Carl George von Staudt (in his *Beiträge zur Geometrie der Lage* of 1847). The cross-ratio concept has become basic in projective geometry, where its power and applicability are of prime importance.

1.5.1 Definition and notation. If A, B, C, D are four distinct points on an ordinary line, we designate the ratio of ratios

$$(\overline{AC}/\overline{CB})/(\overline{AD}/\overline{DB})$$

by the symbol (AB, CD), and call it the *cross ratio* (or *anharmonic ratio*, or *double ratio*) of the range of points A, B, C, D, taken in this order.

Essentially the notation (AB, CD) was introduced by Möbius in 1827. He employed the term *Doppelschnitt-Verhältniss*, and this was later abbreviated by Jacob Steiner to *Doppelverhältniss*, the English equivalent of which is *double ratio*. Chasles used the expression *rapport anharmonique* (*anharmonic ratio*) in 1837, and William Kingdon Clifford coined the term *cross ratio* in 1878. Staudt used the term *Wurf* (*throw*).

The cross ratio of four collinear points depends upon the order in which the points are selected. Since there are twenty-four permutations of four distinct objects, there are twenty-four ways in which a cross ratio of four distinct collinear points may be written. These cross ratios, however, are not all different in value. In fact, we proceed to show that the twenty-four cross ratios may be arranged into six sets of four each, such that the cross ratios in each set have the same value. Indeed, if one of these values is denoted by r, the others are $1/r$, $1 - r$, $1/(1 - r)$, $(r - 1)/r$, and $r/(r - 1)$.

1.5.2 Theorem. *If, in the symbol $(AB, CD) = r$ for the cross ratio of four distinct points, (1) we interchange any two of the points and at the same time interchange the other two points, the cross ratio is unaltered, (2) we interchange only the first pair of points, the resulting cross ratio is $1/r$, (3) we interchange only the middle pair of points, the resulting cross ratio is $1 - r$.*

For the first part we must show that

$$(BA, DC) = (CD, AB) = (DC, BA) = (AB, CD) = r,$$

which is easily accomplished by expanding each of the involved cross ratios.

For the second part we must show that

$$(BA, CD) = 1/r,$$

and this too is easily accomplished by expansion.

For the third part we must show that

$$(AC, BD) = 1 - r.$$

This may be accomplished neatly by dividing the Euler identity (Theorem 1.1.5),

$$\overline{AD} \cdot \overline{BC} + \overline{BD} \cdot \overline{CA} + \overline{CD} \cdot \overline{AB} = 0,$$

by $\overline{AD} \cdot \overline{BC}$, obtaining

$$1 + (\overline{BD} \cdot \overline{CA})/(\overline{AD} \cdot \overline{BC}) + (\overline{CD} \cdot \overline{AB})/(\overline{AD} \cdot \overline{BC}) = 0,$$

or, rearranging,

$$(\overline{AB}/\overline{BC})/(\overline{AD}/\overline{DC}) = 1 - (\overline{AC}/\overline{CB})/(\overline{AD}/\overline{DB}),$$

whence $(AC, BD) = 1 - (AB, CD)$.

1.5.3 THEOREM. *If $(AB, CD) = r$, then*

(1) $(AB, CD) = (BA, DC) = (CD, AB) = (DC, BA) = r$,
(2) $(BA, CD) = (AB, DC) = (DC, AB) = (CD, BA) = 1/r$,
(3) $(AC, BD) = (BD, AC) = (CA, DB) = (DB, CA) = 1 - r$,
(4) $(CA, BD) = (DB, AC) = (AC, DB) = (BD, CA) = 1/(1 - r)$,
(5) $(BC, AD) = (AD, BC) = (DA, CB) = (CB, DA) = (r - 1)/r$,
(6) $(CB, AD) = (DA, BC) = (AD, CB) = (BC, DA) = r/(r - 1)$.

The equalities (1) are guaranteed by the first part of Theorem 1.5.2. The equalities (2) are obtained from those in (1) by applying the operation of the second part of Theorem 1.5.2; the equalities (3) are obtained from those in (1) by applying the operation of the third part of Theorem 1.5.2; the equalities (4) are obtained from those in (3) by applying the operation of the second part of Theorem 1.5.2; the equalities (5) are obtained from those in (2) by applying the operation of the third part of Theorem 1.5.2; the equalities (6) are obtained from those in (5) by applying the operation of the second part of Theorem 1.5.2.

1.5.4 DEFINITION AND NOTATION. If VA, VB, VC, VD are four distinct coplanar lines passing through an ordinary point V, we designate the ratio of ratios

$$(\sin \overline{AVC}/\sin \overline{CVB})/(\sin \overline{AVD}/\sin \overline{DVB})$$

by the symbol $V(AB, CD)$, and call it the *cross ratio* of the pencil of lines VA, VB, VC, VD, taken in this order. It is to be observed (see Theorem 1.1.10) that this definition is independent of the positions of the points A, B, C, D on their respective lines, so long as they are distinct from V.

1.5.5 THEOREM. *If four distinct parallel lines a, b, c, d are cut by two transversals in the points A, B, C, D and A', B', C', D' respectively, then $(AB, CD) = (A'B', C'D')$.*

This follows immediately from the fact that the segments cut off by the parallel lines on one transversal are proportional to the corresponding segments cut off on the other transversal.

1.5.6 DEFINITION. The *cross ratio* of a pencil of four distinct parallel lines, a, b, c, d is taken to be the cross ratio of the range A, B, C, D cut off by the parallel lines on any transversal to these lines.

We now state and prove the theorem which gives to cross ratio its singular power in projective geometry.

1.5.7 THEOREM. *The cross ratio of any pencil of four distinct lines is equal to the cross ratio of the corresponding four points in which any ordinary transversal cuts the pencil.*

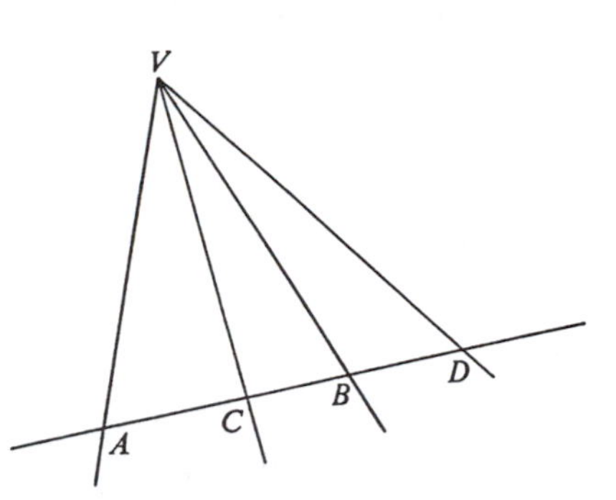

Figure 1.5a

If the vertex of the pencil is a point at infinity, the theorem follows from Definition 1.5.6.

Suppose the vertex V of the pencil is not at infinity, and let A, B, C, D be the points in which the pencil is cut by an ordinary transversal (see Figure 1.5a). Then, by Theorem 1.1.9,

$$\overline{AC}/\overline{CB} = (VA \sin \overline{AVC})/(VB \sin \overline{CVB}),$$
$$\overline{AD}/\overline{DB} = (VA \sin \overline{AVD})/(VB \sin \overline{DVB}),$$

whence

$$(\overline{AC}/\overline{CB})/(\overline{AD}/\overline{DB}) = (\sin \overline{AVC}/\sin \overline{CVB})/(\sin \overline{AVD}/\sin \overline{DVB}).$$

It follows that $(AB, CD) = V(AB, CD)$, and the theorem is established.

PROBLEMS

1. If we extend the definition of the cross ratio of four collinear points so that two, but no more than two, of the four points may coincide, show that given three distinct collinear points A, B, C there exists a unique point D collinear with them such that $(AB, CD) = r$, where r has any real value or is infinite. Also show that two points coincide if and only if $r = 0$, 1, or ∞.

2. If A, B, C, D are four distinct collinear points, show that the pairs A, B and C, D do or do not separate each other according as (AB, CD) is negative or positive.

3. Given three distinct points A, B, C on a line l, construct a fourth point D collinear with them such that (AB, CD) shall have a given value r.

4. If P, Q, R, S, T are collinear points, show that
   (a) $(PQ, RT)(PQ, TS) = (PQ, RS)$.
   (b) $(PT, RS)(TQ, RS) = (PQ, RS)$.

5. If O, A, B, C, A', B', C' are collinear and if $\overline{OA} \cdot \overline{OA'} = \overline{OB} \cdot \overline{OB'} = \overline{OC} \cdot \overline{OC'}$, show that $(AB', BC) = (A'B, B'C')$.

6. If $(AB, CP) = m$ and $(AB, CQ) = n$, show that $(AC, PQ) = (n - 1)/(m - 1)$.

7. Investigate the cases where two of the six values of the cross ratios of four collinear points are equal.

8. Show that the six cross ratios of four collinear points can be represented by $\cos^2 \theta$, $\csc^2 \theta$, $-\tan^2 \theta$, $\sec^2 \theta$, $\sin^2 \theta$, $-\cot^2 \theta$. Show that 2θ is the angle of intersection of the circles described on AB and CD as diameters, it being supposed that the points are in the order A, C, B, D. (This result is due to J. Casey.)

1.6 APPLICATIONS OF CROSS RATIO

We first state a number of useful corollaries to Theorem 1.5.7. The proofs can easily be supplied by the reader.

1.6.1 COROLLARY. *If A, B, C, D and A', B', C', D' are two coplanar ranges on distinct bases such that $(AB, CD) = (A'B', C'D')$, and if AA', BB', CC' are concurrent, then DD' passes through the point of concurrence.*

1.6.2 COROLLARY. *If A, B, C, D and A', B', C', D' are two coplanar ranges on distinct bases such that $(AB, CD) = (A'B', C'D')$, and if A and A' coincide, then BB', CC', DD' are concurrent.*

1.6.3 COROLLARY. *If VA, VB, VC, VD and $V'A$, $V'B$, $V'C$, $V'D$ are two coplanar pencils on distinct vertices V and V' such that $V(AB, CD) = V'(AB, CD)$, and if A, B, C are collinear, then D lies on the line of collinearity.*

1.6.4 COROLLARY. *If VA, VB, VC, VD and $V'A$, $V'B$, $V'C$, $V'D$ are two coplanar pencils on distinct vertices V and V' such that $V(AB, CD) = V'(AB, CD)$, and if A lies on VV', then B, C, D are collinear.*

The next theorem, which is an immediate consequence of the elementary angle relations in a circle, gives an important cross-ratio property of the circle.

1.6.5 THEOREM. *If A, B, C, D are any four distinct points on a circle, and if V and V' are any two points on the circle, then $V(AB, CD) =$*

$V'(AB, CD)$, where VA, say, is taken as the tangent to the circle at A if V should coincide with A.

1.6.6 DEFINITION AND NOTATION. If A, B, C, D are four distinct points on a circle, and V is any fifth point on the circle, we shall designate the cross ratio $V(AB, CD)$, which, by Theorem 1.6.5, is independent of the position of V, by the symbol (AB, CD), and we shall call it the *cross ratio* of the *cyclic range* of points A, B, C, D, taken in this order.

1.6.7 THEOREM. *If A, B, C, D are four distinct points on a circle, then $(AB, CD) = e(AC/CB)/(AD/DB)$, where AC, CB, AD, DB are chord lengths, and where e is -1 or $+1$ according as the pairs A, B and C, D do or do not separate each other.*

For, denoting the center of the circle by O and its radius by r, we have

$$AC = 2r\sin(AOC/2) = 2r\sin AVC, \text{ etc.},$$

whence

$$\begin{aligned}(AB, CD) = V(AB, CD) &= (\sin \overline{AVC}/\sin \overline{CVB})/(\sin \overline{AVD}/\sin \overline{DVB}) \\ &= \pm (AC/CB)/(AD/DB).\end{aligned}$$

The reader can easily show that we must take the minus sign if the pairs of rays VA, VB and VC, VD separate each other, and that otherwise we must take the plus sign.

We now illustrate the power of cross ratio by proving anew Desargues' two-triangle theorem (Theorem 1.4.3) and Pascal's "mystic hexagram" theorem for a circle (Theorem 1.4.4). Further illustrations will be found among the problems at the end of this section and in various other parts of the book.

1.6.8 DESARGUES' TWO-TRIANGLE THEOREM. *Copolar triangles are coaxial, and conversely.*

Let the two triangles (see Figure 1.6a) be ABC and $A'B'C'$. Suppose AA', BB', CC' are concurrent in a point O. Let P, Q, R be the points of intersection of BC and $B'C'$, CA and $C'A'$, AB and $A'B'$. Let CC' cut AB in X, $A'B'$ in X', and PQ in Y. Then, by successive applications of Theorem 1.5.7,

$$\begin{aligned}C(YP, QR) = (XB, AR) = O(XB, AR) &= (X'B', A'R) \\ &= C'(X'B', A'R) = C'(YP, QR).\end{aligned}$$

Since Y lies on CC', it follows (by Corollary 1.6.4) that P, Q, R are collinear.

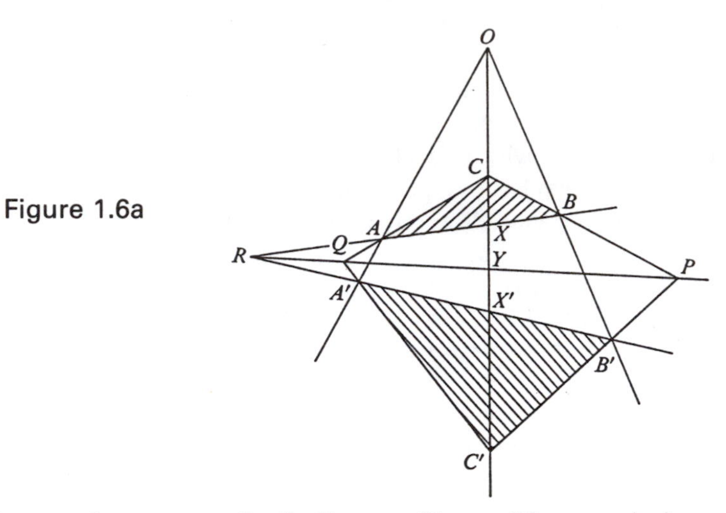

Figure 1.6a

Conversely, suppose P, Q, R are collinear. Then, again by successive applications of Theorem 1.5.7,

$$(RA, XB) = C(RA, XB) = C(RQ, YP) = C'(RQ, YP) = (RA', X'B').$$

It now follows (by Corollary 1.6.2) that AA', XX' (or CC'), BB' are concurrent.

1.6.9 PASCAL'S "MYSTIC HEXAGRAM" THEOREM FOR A CIRCLE. *The points L, M, N of intersection of the three pairs of opposite sides AB and DE, BC and EF, FA and CD of a (not necessarily convex) hexagon $ABCDEF$ inscribed in a circle lie on a line.*

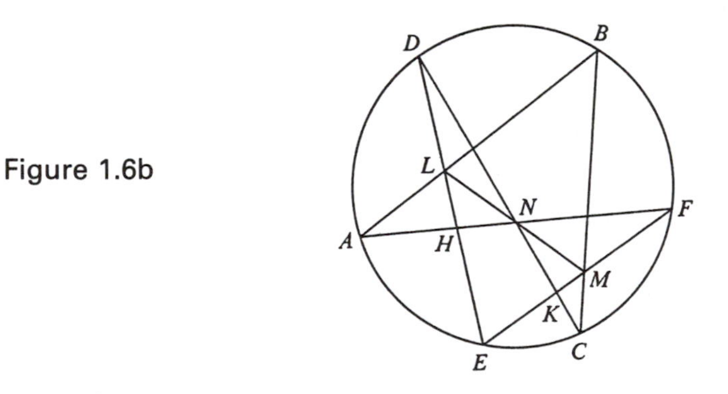

Figure 1.6b

Let AF and ED (see Figure 1.6b) intersect in H, and EF and CD in K. Then (by Theorem 1.6.5) $A(EB, DF) = C(EB, DF)$, whence (by Theorem 1.5.7) $(EL, DH) = (EM, KF)$. It now follows (by Corollary 1.6.2) that LM, DK, HF are concurrent, or that L, M, N are collinear.

PROBLEMS

1. Supply proofs for Corollaries 1.6.1, 1.6.2, 1.6.3, and 1.6.4.

2. Prove Theorem 1.6.5.

3. Let a, b, c, d be four distinct fixed tangents to a given circle and let p be a variable fifth tangent. If p cuts a, b, c, d in A, B, C, D, show that (AB, CD) is a constant independent of the position of p.

4. Prove *Brianchon's Theorem* for a circle: If $ABCDEF$ is a (not necessarily convex) hexagon circumscribed about a circle, then AD, BE, CF are concurrent.

5. If two transversals cut the sides BC, CA, AB of a triangle ABC in points P, Q, R and P', Q', R', show that

$$(BC, PP')(CA, QQ')(AB, RR') = 1.$$

6. If A, B, C, D, A', B', C', D', M, N are collinear, and if $(AA', MN) = (BB', MN) = (CC', MN) = (DD', MN)$, show that $(AB, CD) = (A'B', C'D')$.

7. Prove *Pappus' Theorem*: If A, C, E and B, D, F are two sets of three points on distinct lines, then the points of intersection of AB and DE, BC and EF, FA and CD are collinear.

8. In a hexagon $AC'BA'CB'$, BB', $C'A$, $A'C$ are concurrent and CC', $A'B$, $B'A$ are concurrent. Prove that AA', $B'C$, $C'B$ are also concurrent.

1.7 HOMOGRAPHIC RANGES AND PENCILS (OPTIONAL)

This section is devoted to a brief consideration of homographic ranges and pencils.

1.7.1 DEFINITIONS. If the points of two complete ranges are paired in one-to-one correspondence such that the cross ratio of each four points of one range is equal to the cross ratio of the corresponding four points of the other range, then the two ranges are said to be *homographic* (to each other).

If the lines of two complete pencils are paired in one-to-one correspondence such that the cross ratio of each four lines of one pencil is equal to the cross ratio of the corresponding four lines of the other pencil, then the two pencils are said to be *homographic* (to each other).

1.7.2 THEOREM. (*1*) *If the joins of three distinct pairs of corresponding points of two homographic ranges on distinct bases are concurrent in a point O, then the joins of all pairs of corresponding points of the two ranges pass through O.* (*2*) *If the intersections of three distinct pairs of corresponding lines of two homographic pencils on distinct vertices are* *collinear on a line l, then the intersections of all pairs of corresponding lines of the two pencils lie on l.*

This theorem is an immediate consequence of Corollaries 1.6.1 and 1.6.3.

1.7.3 THEOREM. (*1*) *If two homographic ranges on distinct bases have a pair of corresponding points coincident, then the joins of all other pairs of corresponding points are concurrent.* (*2*) *If two homographic pencils on distinct vertices have a pair of corresponding lines coincident, then the intersections of all other pairs of corresponding lines are collinear.*

This theorem is an immediate consequence of Corollaries 1.6.2 and 1.6.4.

1.7.4 THEOREM. *A necessary and sufficient condition that two complete ranges be homographic is that the points of the two ranges be paired in one-to-one correspondence such that the signed distances x and x' of a variable pair of corresponding points X and X', measured from fixed origins on the bases of the two ranges, be connected by a relation of the form*

$$rxx' + sx + tx' + u = 0,$$

where r, s, t, u are real numbers and $ru - st \neq 0$.

Necessity. Suppose the ranges are homographic. Let A and A', B and B', C and C' be three distinct fixed pairs of corresponding points and X and X' be an arbitrary pair of corresponding points of the two ranges. Then $(AB, CX) = (A'B', C'X')$, or

$$(\overline{AC}/\overline{CB})/(\overline{AX}/\overline{XB}) = (\overline{A'C'}/\overline{C'B'})/(\overline{A'X'}/\overline{X'B'}),$$

or, letting corresponding lower case letters represent the signed distances of the points from the fixed origins,

$$\frac{(c-a)(b-x)}{(b-c)(x-a)} = \frac{(c'-a')(b'-x')}{(b'-c')(x'-a')},$$

which reduces to

$$rxx' + sx + tx' + u = 0,$$

where

$$\begin{aligned} r &= (c'-a')(b-c) - (c-a)(b'-c'), \\ s &= a'(c-a)(b'-c') - b'(c'-a')(b-c), \\ t &= b(c-a)(b'-c') - a(c'-a')(b-c), \end{aligned}$$

$$u = b'a(c' - a')(b - c) - ba'(c - a)(b' - c').$$

A little algebra shows that

$$ru - st = (c - a)(c' - a')(b - c)(b' - c')(a - b)(a' - b'),$$

which is different from zero since A, B, C and A', B', C' are distinct points on their bases.

Sufficiency. Suppose the ranges are in one-to-one correspondence with

$$rxx' + sx + tx' + u = 0, \qquad ru - st \neq 0.$$

Then

$$x' = -(sx + u)/(rx + t).$$

Let A and A', B and B', C and C', D and D' be any four distinct pairs of corresponding points. Now

$$(A'B', C'D') = \frac{\overline{A'C'}/\overline{C'B'}}{\overline{A'D'}/\overline{D'B'}} = \frac{(c' - a')(b' - d')}{(b' - c')(d' - a')}.$$

But

$$c' - a' = -\frac{sc + u}{rc + t} + \frac{sa + u}{ra + t} = \frac{(ru - st)(c - a)}{(rc + t)(ra + t)},$$

with similar expressions for $b' - d'$, $b' - c'$, $d' - a'$. Therefore

$$\frac{(c' - a')(b' - d')}{(b' - c')(d' - a')} = \frac{(c - a)(b - d)}{(b - c)(d - a)},$$

since all other factors cancel. That is, $(A'B', C'D') = (AB, CD)$.

There are two classes of problems for which Theorem 1.7.3 is often applicable, namely: (1) those in which it is required to prove that the locus of a variable point is a straight line, (2) those in which it is required to prove that a variable straight line passes through a fixed point. In (1) we obtain two homographic pencils having a common line, namely the line joining the vertices of the pencils, and having the different positions of the variable point as the intersections of pairs of corresponding lines of the pencils. In (2) we obtain two homographic ranges having a common point, namely the point of intersection of the bases of the ranges, and having the different positions of the variable line as the joins of pairs of corresponding points of the ranges. Some problems of this sort are found among those appearing at the end of the section. Theorem 1.7.4 often furnishes an easy way of showing that two complete ranges are homographic.

PROBLEMS

1. Prove that two ranges (pencils) which are homographic to the same range (pencil) are homographic to each other.

2. A and B are two fixed points and X and X' are two variable points on two given lines intersecting in a point O. If $\overline{OX} + \overline{OX'} = \overline{OA} + \overline{OB}$, show that the locus of the point of intersection of AX' and BX is a straight line.

3. A variable triangle ABC has sides BC, CA, AB which pass through fixed points P, Q, R respectively. If the vertices B and C move along given lines through a point O collinear with Q and R, find the locus of vertex A.

4. A variable triangle ABC has vertices A, B, C which move along fixed lines p, q, r respectively. If the sides CA and AB pass through given points on a line l concurrent with q and r, show that side BC passes through a fixed point.

5. Three points F, G, H are taken on side BC of a triangle ABC. Through G a variable line is drawn cutting AB and AC in L and M respectively. Prove that the intersection K of FL and HM is a straight line passing through A.

6. A point P, which moves along a given straight line, is joined to two fixed points B and C, and the lines PB, PC cut another line in X and Y. Find the locus of the point of intersection of BY and CX.

7. A, D, C are three fixed collinear points, E is a fixed point not on line ADC, B is a variable point on line CE. The lines AE and BD intersect in Q, CQ and DE intersect in R, BR and AC intersect in P. Show that P is a fixed point.

8. Let (A) and (A') be two homographic ranges. If I corresponds to the point at infinity on (A') and J' to the point at infinity on (A), show that $(\overline{IX})(\overline{J'X'})$ is constant for all pairs of corresponding points X and X'.

9. If I and J' are fixed points, and X and X' are variable points, on two straight lines, and if $(\overline{IX})(\overline{J'X'}) =$ constant, show that X and X' generate homographic ranges in which I and J' are the points corresponding to the points at infinity on the two straight lines.

10. (a) Let a variable circle through a point V and cutting a line l not through V in a given angle θ cut l in points P and P', P lying to the left of P' when l is considered as horizontal. Show that (P) and (P') are homographic ranges.
    (b) Let $J'VI$ be an isosceles triangle with base lying on l and having base angles θ, J' lying to the left of I. Show that $(\overline{IP})(\overline{J'P'}) =$ constant.

11. If the vertices of a polygon move on fixed concurrent lines, and all but one of the sides pass through fixed points, show that this side and each diagonal will pass through fixed points.

12. If each side of a polygon passes through one of a set of collinear points while all but one of its vertices slide on fixed lines, show that the remaining vertex and each intersection of two sides will describe lines.

13. Let four lines through a point V cut a circle in A, A'; B, B'; C, C'; D, D' respectively. Show that $(AB, CD) = (A'B', C'D')$.

1.8 HARMONIC DIVISION

Very important and particularly useful is the special cross ratio having the value -1. Since, as we shall see, there is an intimate connection between such a cross ratio and three numbers in harmonic progression, such a cross ratio is referred to as a "harmonic division."

1.8.1 DEFINITIONS. If A, B, C, D are four collinear points such that $(AB, CD) = -1$ (so that C and D divide AB, one internally and the other externally, in the same numerical ratio), the segment AB is said to be *divided harmonically* by C and D, the points C and D are called *harmonic conjugates* of each other with respect to A and B, and the four points A, B, C, D are said to constitute a *harmonic range*. If $V(AB, CD) = -1$, then VA, VB, VC, VD are said to constitute a *harmonic pencil*.

1.8.2 THEOREM. *If C and D divide AB harmonically, then A and B divide CD harmonically.*

For if $(AB, CD) = -1$, then (by Theorem 1.5.2(1)) so also does $(CD, AB) = -1$.

1.8.3 THEOREM. *The harmonic conjugate with respect to A and B of the midpoint of AB is the point at infinity on AB.*

Let M be the midpoint of AB. Then $\overline{AM}/\overline{MB} = 1$. It follows that if D is the harmonic conjugate of M with respect to A and B we must have $\overline{AD}/\overline{DB} = -1$. Thus (see Definition 1.2.3) D is the point at infinity on line AB.

The next two theorems furnish useful criteria for four collinear points to form a harmonic range.

1.8.4 THEOREM. *$(AB, CD) = -1$ if and only if $2/\overline{AB} = 1/\overline{AC} + 1/\overline{AD}$.*

Suppose $(AB, CD) = -1$. Then $\overline{AC}/\overline{CB} = -\overline{AD}/\overline{DB}$, whence

$$\overline{CB}/(\overline{AB} \cdot \overline{AC}) = \overline{BD}/(\overline{AB} \cdot \overline{AD}),$$

or

$$(\overline{AB} - \overline{AC})/(\overline{AB} \cdot \overline{AC}) = (\overline{AD} - \overline{AB})/(\overline{AB} \cdot \overline{AD}).$$

That is,

$$1/\overline{AC} - 1/\overline{AB} = 1/\overline{AB} - 1/\overline{AD},$$

or

$$2/\overline{AB} = 1/\overline{AC} + 1/\overline{AD}.$$

The converse may be established by reversing the above steps.

1.8.5 THEOREM. *$(AB, CD) = -1$ if and only if $\overline{OB}^2 = \overline{OC} \cdot \overline{OD}$, where O is the midpoint of AB.*

Suppose $(AB, CD) = -1$. Then $\overline{AC}/\overline{CB} = -\overline{AD}/\overline{DB}$, whence

$$(\overline{OC} - \overline{OA})/(\overline{OB} - \overline{OC}) = -(\overline{OD} - \overline{OA})/(\overline{OB} - \overline{OD}),$$

or, since $\overline{OA} = -\overline{OB}$,

$$(\overline{OC} + \overline{OB})/(\overline{OB} - \overline{OC}) = (\overline{OD} + \overline{OB})/(\overline{OD} - \overline{OB}).$$

It now follows that

$$(\overline{OC} + \overline{OB})(\overline{OD} - \overline{OB}) = (\overline{OD} + \overline{OB})(\overline{OB} - \overline{OC}),$$

or, upon multiplying out and simplifying,

$$\overline{OB}^2 = \overline{OC} \cdot \overline{OD}.$$

The converse may be established by reversing the above steps.

We now look at the connection between "harmonic division" and "harmonic progression."

1.8.6 DEFINITION. The sequence of numbers $\{a_1, a_2, \ldots, a_n\}$ is said to be a *harmonic progression* if the sequence of numbers $\{1/a_1, 1/a_2, \ldots, 1/a_n\}$ is an arithmetic progression.

1.8.7 THEOREM. *The sequence of numbers $\{a_1, a_2, a_3\}$ is a harmonic progression if and only if $2/a_2 = 1/a_1 + 1/a_3$.*

The proof follows readily from Definition 1.8.6.

1.8.8 THEOREM. *If $(AB, CD) = -1$, then $\{\overline{AC}, \overline{AB}, \overline{AD}\}$ is a harmonic progression.*

This is a consequence of Theorems 1.8.4 and 1.8.7.

We conclude this section with a brief consideration of complete quadrilaterals and complete quadrangles, and their useful harmonic properties.

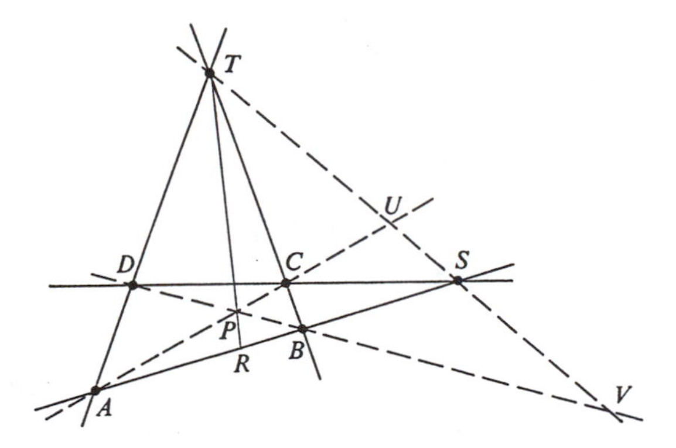

Figure 1.8a

1.8.9 DEFINITIONS. A *complete quadrilateral* (see Figure 1.8a) is the figure formed by four coplanar lines, no three of which are concurrent. The four lines are called the *sides* of the complete quadrilateral, and the six points of intersection of pairs of the sides are called the *vertices* of the complete quadrilateral. Pairs of vertices not lying on any common side are called *opposite vertices* of the complete quadrilateral. The lines through the three pairs of opposite vertices are called the *diagonal lines* of the complete quadrilateral, and the triangle determined by the three diagonal lines is called the *diagonal 3-line* of the complete quadrilateral.

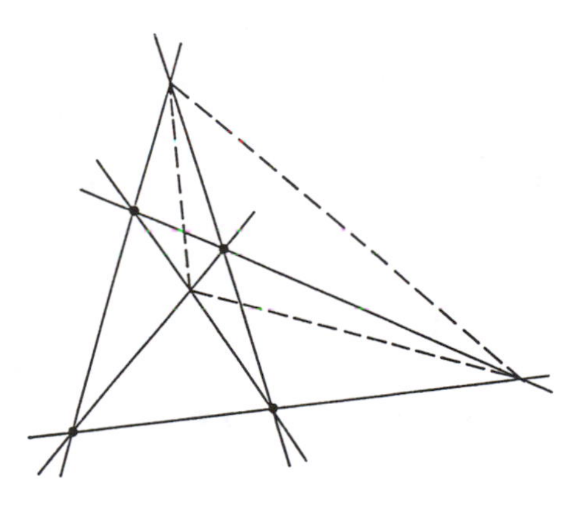

Figure 1.8b

A *complete quadrangle* (see Figure 1.8b) is the figure formed by four coplanar points, no three of which are collinear. The four points are called the *vertices* of the complete quadrangle, and the six lines determined by pairs of the vertices are called the *sides* of the complete quadrangle. Pairs of sides not passing through any common vertex are called *opposite sides* of the complete quadrangle. The points of intersection of the three pairs of opposite sides are called the *diagonal points* of the complete quadrangle, and the triangle determined by the three diagonal points is called the *diagonal 3-point* of the complete quadrangle.

1.8.10 THEOREM. *On each diagonal line of a complete quadrilateral there is a harmonic range consisting of the two vertices of the quadrilateral and the two vertices of the diagonal 3-line lying on it.*

Let us show (see Figure 1.8a) that $(UV, TS) = -1$. We have

$$(UV, TS) = P(UV, TS) = P(AB, RS) = (AB, RS).$$

But, by Theorem 1.4.1, $(AB, RS) = -1$. We may now easily show that $(PU, AC) = (PV, DB) = (UV, TS) = -1$.

1.8.11 THEOREM. *At each diagonal point of a complete quadrangle there is a harmonic pencil consisting of the two sides of the quadrangle and the two sides of the diagonal 3-point passing through it.*

This follows immediately from Theorem 1.4.1.

PROBLEMS

1. Justify each of the following methods of constructing the harmonic conjugate D of a given point C with respect to a given segment AB.
   (a) Take any point P not on line AB and connect P to A, B, C. Through B draw the parallel to AP, cutting line PC in M, and on this parallel mark off $\overline{BN} = \overline{MB}$. Then PN cuts line AB in the sought point D.
   (b) Draw the circle on AB as diameter. If C lies between A and B, draw CT perpendicular to AB to cut the circle in T. Then the tangent to the circle at T cuts line AB in the sought point D. If C is not between A and B, draw one of the tangents CT to the circle, T being the point of contact of the tangent. Then the sought point D is the foot of the perpendicular dropped from T on AB.
   (c) Connect any point P not on line AB with A, B, C. Through A draw any line (other than AB or AP) to cut PC and PB in M and N respectively. Draw BM to cut PA in G. Now draw GN to cut line AB in the sought point D. (Note that this construction uses only a straightedge.)

2. If $(AB, CD) = -1$ and O and O' are the midpoints of AB and CD respectively, show that $(OB)^2 + (O'C)^2 = (OO')^2$.

3. (a) Show that the lines joining any point on a circle to the vertices of an inscribed square form a harmonic pencil.
   (b) Show, more generally, that the lines joining any point on a circle to the extremities of a given diameter and to the extremities of a given chord perpendicular to the diameter form a harmonic pencil.
   (c) Triangle ABC is inscribed in a circle of which DE is the diameter perpendicular to side AC. If lines DB and EB intersect AC in L and M, show that $(AC, LM) = -1$.
   (d) Show that the diameter of a circle perpendicular to one of the sides of an inscribed triangle is divided harmonically by the other two sides.

4. (a) If L, M, N are the midpoints of the sides BC, CA, AB of a triangle ABC, show that $L(MN, AB) = -1$.
(b) If P, Q, R are the feet of the altitudes on sides BC, CA, AB of a triangle ABC, show that $P(QR, AB) = -1$.
(c) The bisector of angle A of triangle ABC intersects the opposite side in T. U and V are the feet of the perpendiculars from B and C upon line AT. Show that $(AT, UV) = -1$.

5. Let BC be a diameter of a given circle, let A be a point on BC produced, and let P and Q be the points of contact of the tangent to the circle from point A. Show that $P(AQ, CB) = -1$.

6. If $P(AB, CD) = -1$ and if PC is perpendicular to PD, show that PC and PD are the bisectors of angle APB.

7. If O is any point on the altitude AP of triangle ABC, and BO and CO intersect AC and AB in E and F respectively, show that PA bisects angle EPF.

8. Given four collinear points A, B, C, D, find points P and Q such that $(AB, PQ) = (CD, PQ) = -1$.

9. In triangle ABC we have $(BC, PP') = (CA, QQ') = (AB, RR') = -1$. Show that AP', BQ', CR' are concurrent if and only if P, Q, R are collinear.

10. Prove, in Figure 1.8a, that

$$TA \cdot TC \cdot SB \cdot SD = SA \cdot SC \cdot TB \cdot TD.$$

11. If $(AB, CD) = -1$ and O is the midpoint of CD, show that $\overline{AC} \cdot \overline{AD} = \overline{AB} \cdot \overline{AO}$.

12. If P, P' divide one diameter of a circle harmonically and Q, Q' divide another diameter harmonically, prove that P, Q, P', Q' are concyclic.

13. Two circles intersect in points A and B. A common tangent touches the circles at P and Q and cuts a third circle through A and B in L and M. Prove that $(PQ, LM) = -1$.

14. A circle Σ inscribed in a semicircle touches the diameter AB of the semicircle at a point C. Prove that the diameter of Σ is the harmonic mean between AC and CB.

15. In Figure 1.8a, prove that UD, VA, PS are concurrent.

16. If $(AB, CD) = -1$ and O is the midpoint of AB, prove that $\overline{OC}/\overline{OD} = (AC)^2/(AD)^2$.

17. In Figure 1.8a, prove that the midpoints X, Y, Z of DB, AC, ST are collinear.

18. A secant from an external point A cuts a circle in C and D, and cuts the chord of contact of the tangents to the circle from A in B. Prove that $(AB, CD) = -1$.

19. If A, B, C are collinear, P the harmonic conjugate of A with respect to B and

C, Q the harmonic conjugate of B with respect to C and A, and R the harmonic conjugate of C with respect to A and B, show that A is the harmonic conjugate of P with respect to Q and R.

20. If P', Q' are the harmonic conjugates of P and Q with respect to A and B, show that the segments PQ, $P'Q'$ subtend equal or supplementary angles at any point on the circle described on AB as diameter.

21. Prove that the geometric mean of two positive numbers is the geometric mean of the arithmetic mean and harmonic mean of the two numbers.

22. If $(AB, PQ) = -1$ and O is collinear with A, B, P, Q, show that $2(\overline{OB}/\overline{AB}) = \overline{OP}/\overline{AP} + \overline{OQ}/\overline{AQ}$.

23. In triangle ABC, D and D', E and E', F and F' are harmonic conjugates with respect to B and C, C and A, A and B respectively. Prove that corresponding sides of triangles DEF, $D'E'F'$ intersect on the sides of triangle ABC.

24. Through a given point O a variable line is drawn cutting two fixed lines in P and Q, and on OPQ point X is taken such that $1/\overline{OX} = 1/\overline{OP} + 1/\overline{OQ}$. Find the locus of X.

25. Through a given point O a variable line is drawn cutting n fixed lines in P_1, $P_2, \ldots, P_n$, and on the variable line a point X is taken such that $1/\overline{OX} = 1/\overline{OP_1} + 1/\overline{OP_2} + \cdots + 1/\overline{OP_n}$. Find the locus of X.

1.9 ORTHOGONAL CIRCLES

For later purposes we shall need certain parts of the elementary geometry of circles that did not make their appearance until the nineteenth century. The material we are concerned about is that centered around the concepts of orthogonal circles, the power of a point with respect to a circle, the radical axis of a pair of circles, the radical center and radical circle of a trio of circles, and coaxial pencils of circles. Though one can see the notion of power of a point with respect to a circle foreshadowed in Propositions 35 and 36 of Book III of Euclid's *Elements*, the concept was first crystallized and developed by Louis Gaultier in a paper published in 1813 in the *Journal de l'École Polytechnique*. Here we find, for the first time, the terms *radical axis* and *radical center*; the term *power* was introduced somewhat later by Jacob Steiner. The initial studies of orthogonal circles and coaxial pencils of circles were made in the early nineteenth century by Gaultier, Poncelet, Steiner, J. B. Durrende, and others. This rather recent elementary geometry of the circle has found valuable application in various parts of mathematics and physics.

We start with a definition of orthogonal curves.

1.9.1 Definitions. By the *angles of intersection* of two coplanar curves at a point which they have in common is meant the angles between the tangents to the curves at the common point. If the angles of intersection are right angles, the two curves are said to be *orthogonal*.

The facts stated about circles in the following theorem are quite obvious.

1.9.2 THEOREM. *(1) The angles of intersection at one of the common points of two intersecting circles are equal to those at the other common point. (2) If two circles are orthogonal, a radius of either, drawn to a point of intersection, is tangent to the other; conversely, if the radius of one of two intersecting circles, drawn to a point of intersection, is tangent to the other, the circles are orthogonal. (3) Two circles are orthogonal if and only if the square of the distance between their centers is equal to the sum of the squares of their radii. (4) If two circles are orthogonal, the center of each lies outside the other.*

We now establish a few deeper facts about orthogonal circles that will be useful to us in a later chapter.

1.9.3 THEOREM. *If two circles are orthogonal, then any diameter of one which intersects the other is cut harmonically by the other; conversely, if a diameter of one circle is cut harmonically by a second circle, then the two circles are orthogonal.*

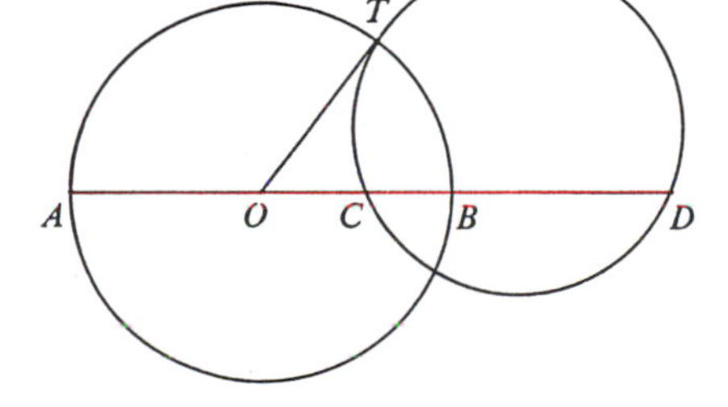

Figure 1.9a

Let O (see Figure 1.9a) be the center of one of a pair of orthogonal circles and let a diameter AOB of this circle cut the other in points C and D. Let T be a point of intersection of the two circles. Then $(OB)^2 = (OT)^2 = (\overline{OC})(\overline{OD})$, since (by Theorem 1.9.2 (2)) OT is tangent to the second circle. It now follows (by Theorem 1.8.5) that $(AB, CD) = -1$.

Conversely, if $(AB, CD) = -1$, then (by Theorem 1.8.5) $(OT)^2 = (OB)^2 = (\overline{OC})(\overline{OD})$, and OT is tangent to the second circle, whence (by Theorem, 1.9.2 (2)) the two circles are orthogonal.

1.9.4 DEFINITION AND CONVENTION. We shall call a circle or a straight line a "*circle*" (with quotation marks), and we shall adopt the convention that two straight lines are *tangent* if and only if they either coincide or are parallel. With this convention, it is perfectly clear in all cases what is meant by two "circles" being tangent to one another. Some authors refer to a "circle" as a *stircle*.

1.9.5 THEOREM. *There is one and only one "circle" orthogonal to a given circle Σ and passing through two given interior points A and B of Σ.*

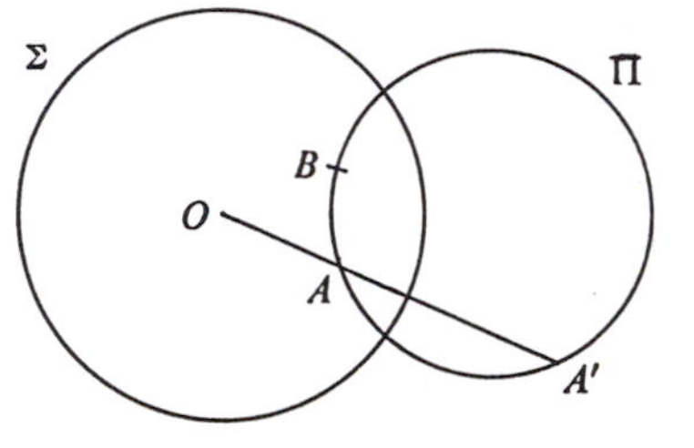

Figure 1.9b

Let O (see Figure 1.9b) be the center of Σ. If A, O, B are collinear, the diameter AOB is a straight line through A and B and orthogonal to Σ. If A, O, B are not collinear, let A' be the harmonic conjugate of A with respect to the endpoints of the diameter of Σ passing through A. Then the circle BAA' passes through A and B and (by Theorem 1.9.3) is orthogonal to Σ. Thus in any event, there is at least one "circle" through the points A and B and orthogonal to Σ. To show that there is only one such "circle," let Π represent any "circle" through A and B and orthogonal to Σ. If Π is a straight line then it must be a diameter of Σ. That is, A, O, B are collinear and Π coincides with the straight line considered earlier. If Π is a circle, then (by Theorem 1.9.3) it must also pass through the point A', and thus coincide with the circle BAA' considered earlier. This proves the theorem.

1.9.6 THEOREM. *There is a unique "circle" orthogonal to a given circle and tangent to a given line l at an ordinary point A of l not on Σ.*

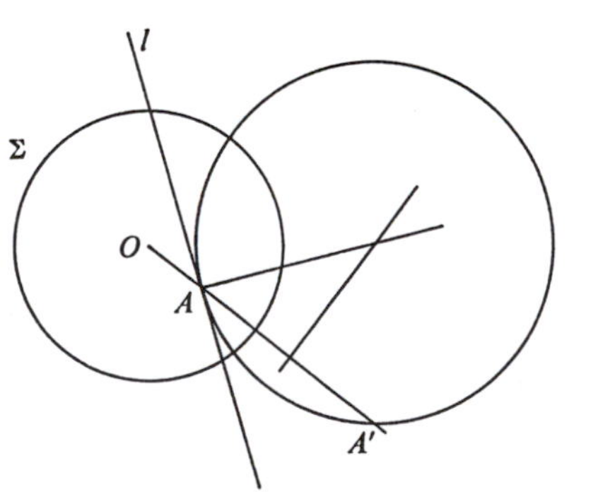

Figure 1.9c

Let O (see Figure 1.9c) be the center of Σ. It is easy to show that if O lies on l, then l is the unique "circle" satisfying the given conditions. If O does not lie on l, it is easy to show that the circle passing through A and having its center at the point of intersection of the perpendicular to l at A and the perpendicular bisector of AA', where A' is the harmonic conjugate of A with respect to the endpoints of the diameter of Σ through A, is the unique "circle" satisfying the given conditions.

As mentioned earlier, Propositions III 35 and III 36 of Euclid's *Elements* contain the germs of the notion of power of a point with respect to a circle. With the aid of sensed magnitudes, these two propositions can be combined into the following single statement.

1.9.7 THEOREM. *If P is a fixed point in the plane of a given circle Σ, and if a variable line l through P intersects Σ in points A and B, then the product $\overline{PA} \cdot \overline{PB}$ is independent of the position of l.*

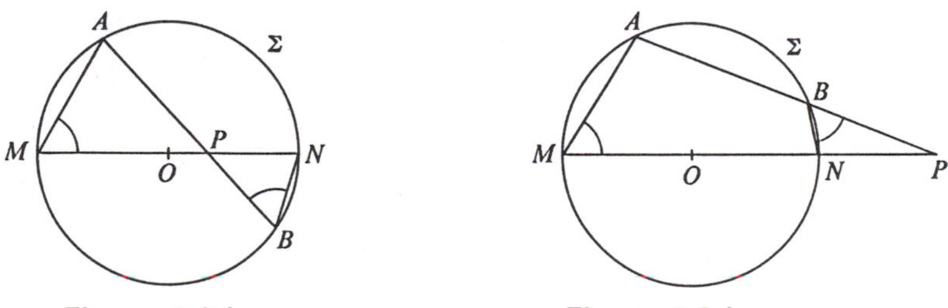

Figure 1.9d$_1$ Figure 1.9d$_2$

Let O be the center of Σ. If P coincides with O, or lies on Σ, or is an ideal point, the theorem is obvious. Otherwise (see Figures 1.9d$_1$ and 1.9d$_2$), draw the diameter MN through the point P and connect A with M and B with N. The two triangles PMA and PBN are equiangular, and therefore similar, whence

$$\overline{PA}/\overline{PM} = \overline{PN}/\overline{PB} \quad \text{or} \quad \overline{PA} \cdot \overline{PB} = \overline{PM} \cdot \overline{PN},$$

and the theorem follows since the right-hand side of the last equality is independent of the position of l.

The preceding theorem justifies the following definition.

1.9.8 DEFINITION. The *power of a point with respect to a circle* is the product of the signed distances of the point from any two points on the circle and collinear with it.

It follows that the power of a point with respect to a circle is positive, zero, or negative according as the point lies outside, on, or inside the circle. If the point lies outside the circle, its power with respect to the circle is equal to the square of the tangent from the point to the circle; if the point lies inside the circle, its power with respect to the circle is the negative of the square of half the chord perpendicular to the diameter passing through the given point. We thus have

1.9.9 THEOREM. *Let P be a point in the plane of a circle Σ of center O and radius r. Then the power of P with respect to Σ is equal to $(OP)^2 - r^2$.*

We leave it to the reader to show that Theorem 1.9.2(3) can be rephrased as follows:

1.9.10 THEOREM. *A necessary and sufficient condition for two circles to be orthogonal is that the power of the center of either with respect to the other be equal to the square of the corresponding radius.*

In 1917, in the *Tôhoku Mathematics Journal* (vol. 11, p. 55), K. Yanagihara carried out an interesting investigation involving the concept of power of a point. He defined a point in the interior of a region bounded by a closed convex curve to be a *power point* for the curve if the product of the segments of a variable chord through the point is a constant. From our work above, it follows that every point in the interior of a circle is a power point for the circle. One naturally wonders what is the minimum number of distinct power points whose existence within a closed convex curve will ensure that the curve be a circle. Since it is easy to construct noncircular closed convex curves possessing only one power point, this minimum number must be greater than 1. Yanagihara showed that if the closed convex curve has a unique tangent at each of its points and possesses two distinct power points, then the curve is a circle. He extended his investigation to 3-space, and showed that if a closed convex surface S having a unique tangent plane at each of its points possesses two power points, then S is a sphere.

PROBLEMS

1. Establish Theorem 1.9.2.
2. Establish Theorem 1.9.10.
3. Show that if d is the distance between the centers of two intersecting circles, c is the length of their common chord, r and r' their radii, then the circles are orthogonal if and only if $cd = 2rr'$.
4. If a line drawn through a point of intersection of two circles meets the circles again in P and Q respectively, show that the circles with centers P and Q, each orthogonal to the other circle, are orthogonal to each other.

5. If AB is a diameter of a circle and if any two lines AC and BC meet the circle again at P and Q respectively, show that circle CPQ is orthogonal to the given circle.

6. Two orthogonal circles intersect in points P and Q. If C is a point on one of the circles, and if CP and CQ cut the other circle in A and B, prove that AB is a diameter of this other circle.

7. Let H be the orthocenter of a triangle ABC. Show that the circles on AH and BC as diameters are orthogonal.

8. If the quadrilateral whose vertices are the centers and the points of intersection of two circles is cyclic, prove that the circles are orthogonal.

9. If AB is a diameter and M any point of a circle of center O, show that the two circles AMO and BMO are orthogonal.

10. Let ABC be a triangle having altitudes AD, BE, CF and orthocenter H. Circles are drawn having A, B, C for centers and $(AH)(AD)$, $(BH)(BE)$, $(CH)(CF)$ for the squares of their respective radii. Prove that each circle is orthogonal to the other two.

1.10 THE RADICAL AXIS OF A PAIR OF CIRCLES

In this section we continue our study of some of the modern elementary geometry of the circle by considering the locus of a point that has equal powers with respect to two given circles. This leads to the important concept of a coaxial pencil of circles.

1.10.1 DEFINITION. The locus of a point whose powers with respect to two given circles are equal is called the *radical axis* of the two given circles.

1.10.2 THEOREM. *The radical axis of two nonconcentric circles is a straight line perpendicular to the line of centers of the two circles.*

Consider two nonconcentric circles with centers O and O' and radii r and r' (see Figure 1.10a), and let P be any point on the radical axis of the two circles. Let Q be the foot of the perpendicular from P on OO'. Then (by Theorem 1.9.9)

$$(PO)^2 - r^2 = (PO')^2 - r'^2.$$

Subtracting $(PQ)^2$ from each side we get

$$(OQ)^2 - r^2 = (QO')^2 - r'^2,$$

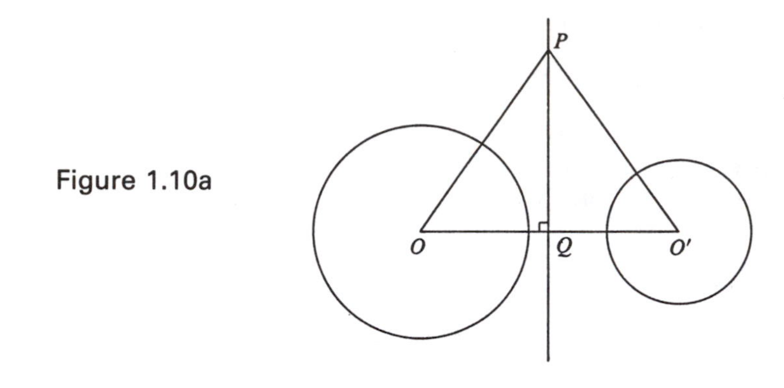

Figure 1.10a

or

$$(\overline{OQ} + \overline{QO'})(\overline{OQ} - \overline{QO'}) = r^2 - r'^2,$$

whence

$$\overline{OQ} - \overline{QO'} = (r^2 - r'^2)/\overline{OO'}. \tag{1}$$

Now there is only one point Q on OO' satisfying relation (1). For if R is any such point we have

$$\overline{OQ} - \overline{QO'} = \overline{OR} - \overline{RO'},$$

or

$$(\overline{OR} + \overline{RQ}) - \overline{QO'} = \overline{OR} - (\overline{RQ} + \overline{QO'}),$$

or

$$\overline{OR} - \overline{QR} - \overline{QO'} = \overline{OR} + \overline{QR} - \overline{QO'},$$

and $\overline{QR} = 0$, or R coincides with Q. It follows that if a point is on the radical axis of the two circles, it lies on the perpendicular to the line of centers at the point Q. Conversely, by reversing the above steps, it can be shown that any point on the perpendicular to OO' at Q lies on the radical axis of the two circles. Therefore the radical axis of the two circles is the perpendicular to OO' at the point Q.

1.10.3 REMARK. If, in equation (1) of the proof of Theorem 1.10.2, $r' \neq r$ and O' approaches O, Q approaches an ideal point. The radical axis of two unequal concentric circles is therefore frequently defined to be the line at infinity in the plane of the circles. The radical axis of two equal concentric circles is left undefined, and it is to be understood that any statement about radical axes is not intended to include this situation.

1.10.4 THEOREM. *The radical axes of three circles with noncollinear centers, taken in pairs, are concurrent.*

Let P be the intersection of the radical axis of the first and second circles with that of the second and third circles. Then P has equal powers with respect to all three circles, and thus must also lie on the radical axis of the first and third circles.

1.10.5 Definition. The point of concurrence of the radical axes of three circles with noncollinear centers, taken in pairs, is called the *radical center* of the three circles.

1.10.6 Theorem. (*1*) *The center of a circle that cuts each of two circles orthogonally lies on the radical axis of the two circles.* (*2*) *If a circle whose center lies on the radical axis of two circles is orthogonal to one of them, it is also orthogonal to the other.*

This is an immediate consequence of Theorem 1.9.10.

1.10.7 Theorem. (*1*) *All the circles that cut each of two given nonintersecting circles orthogonally intersect the line of centers of the two given circles in the same two points.* (*2*) *A circle that cuts each of two given intersecting circles orthogonally does not intersect the line of centers of the two given circles.*

Figure $1.10b_1$

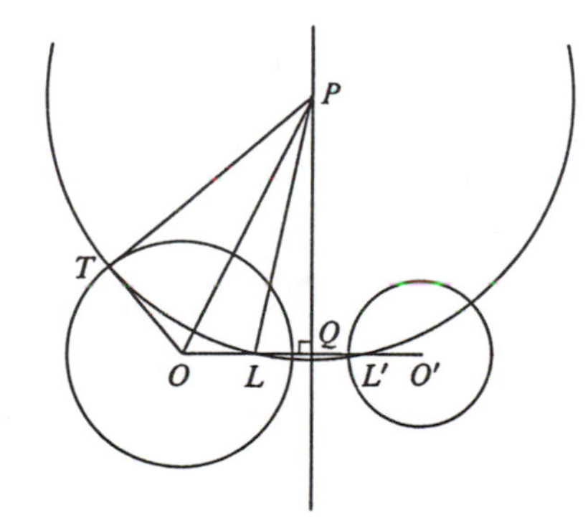

(1) Let a circle with center P cut two given circles with centers O and O' orthogonally. Then (by Theorem 1.10.6) P lies on the radical axis of the two given circles. Referring to Figure $1.10b_1$, we then have $OQ > OT$, whence $PT > PQ$, and the common orthogonal circle intersects OO' in points L and L'. Now

$$(PL)^2 = (LQ)^2 + (QP)^2$$

and also

$$(PL)^2 = (PT)^2 = (PO)^2 - (OT)^2 = (OQ)^2 + (QP)^2 - (OT)^2,$$

whence

$$(LQ)^2 = (OQ)^2 - (OT)^2.$$

This last equation shows that the position of L is independent of that of P. Hence every circle orthogonal to the two given circles passes through point L. Similarly, every such circle passes through L'.

Figure 1.10b$_2$

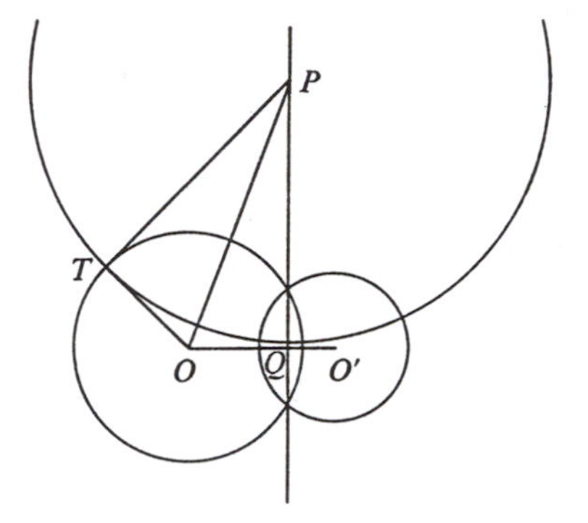

(2) Referring to Figure 1.10b$_2$ we have $OQ < OT$, whence $PT < PQ$, and the common orthogonal circle fails to intersect OO'.

1.10.8 Definitions. A set of circles is said to form a *coaxial pencil* if the same straight line is the radical axis of any two circles of the set; the straight line is called the *radical axis* of the coaxial pencil.

Coaxial pencils of circles are very useful in certain mathematical and physical investigations. We leave to the reader the easy task of establishing the two following important theorems about such sets of circles.

1.10.9 Theorem. (*1*) *The centers of the circles of a coaxial pencil are collinear.* (*2*) *If two circles of a coaxial pencil intersect, every circle of the pencil passes through the same two points; if two circles of a coaxial pencil are tangent at a point, all circles of the pencil are tangent to one another at the same point; if two circles of a coaxial pencil do not intersect, no two circles of the pencil intersect.* (*3*) *The radical axis of a coaxial pencil of circles is the locus of a point whose powers with respect to all the circles of the pencil are equal.*

1.10.10 Definition. By Theorem 1.10.9(2) there are three types of coaxial pencils of circles, and these are called an *intersecting coaxial pencil*, a *tangent coaxial pencil*, and a *nonintersecting coaxial pencil*.

1.10.11 Theorem. (*1*) *All the circles orthogonal to two given nonintersecting circles belong to an intersecting coaxial pencil whose line of centers is the radical axis of the two given circles.* (*2*) *All the circles* *orthogonal to two given tangent circles belong to a tangent coaxial pencil whose line of centers is the common tangent to the two given circles.* (3) *All the circles orthogonal to two given intersecting circles belong to a nonintersecting coaxial pencil whose line of centers is the line of the common chord of the two given circles.*

We close the section with a particularly pretty application of some of the preceding theory to the complete quadrilateral.

1.10.12 THEOREM. *The three circles on the diagonals of a complete quadrilateral as diameters are coaxial; the orthocenters of the four triangles determined by the four sides of the quadrilateral taken three at a time are collinear; the midpoints of the three diagonals are collinear on a line perpendicular to the line of collinearity of the four orthocenters.*

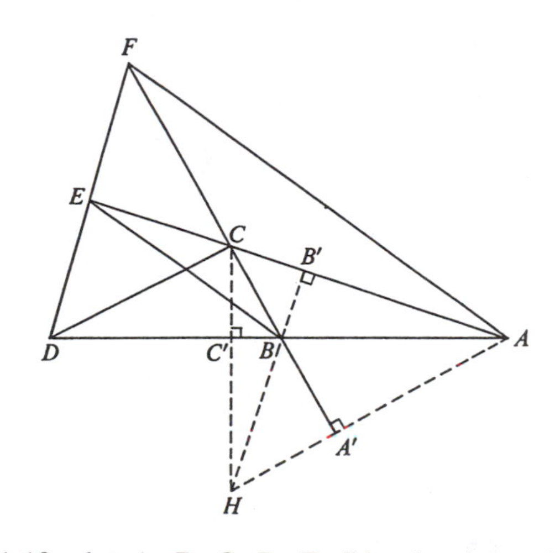

Figure 1.10c

Referring to Figure 1.10c, let A, B, C, D, E, F be the six vertices of a complete quadrilateral, H the orthocenter of triangle ABC, and A', B', C' the feet of the altitudes of triangle ABC. Since A, C, C', A' and B', C, C', B are sets of concyclic points,

$$(\overline{HA})(\overline{HA'}) = (\overline{HB})(\overline{HB'}) = (\overline{HC})(\overline{HC'}).$$

But AA', BB', CC' are chords of the circles having the diagonals AF, BE, CD of the complete quadrilateral as diameters. It follows that H has the same power with respect to all three of these circles. Similarly it can be shown that the orthocenters of triangles ADE, BDF, CEF each have equal powers with respect to the three circles. It follows that the three circles are coaxial, the four orthocenters are collinear on their radical axis, and the centers of the circles (that is, the midpoints of the three diagonals) are collinear on a line perpendicular to the line of collinearity of the four orthocenters.

PROBLEMS

1. Show that if the radical center of three circles with noncollinear centers is exterior to each of the three circles, it is the center of a circle orthogonal to all three circles. (This circle is called the *radical circle* of the three circles.)

2. Establish Theorem 1.10.9.

3. Establish Theorem 1.10.11.

4. Prove that the radical axis of two circles having a common tangent bisects the segment on the common tangent determined by the points of contact.

5. Justify the following construction of the radical axis of two nonconcentric nonintersecting circles. Draw any circle cutting the given circles in A, A' and B, B' respectively. Through P, the intersection of AA' and BB', draw the perpendicular to the line of centers of the given circles. This perpendicular is the required radical axis.

6. (a) Prove that the radical center of the three circles constructed on the sides of a triangle as diameters is the orthocenter of the triangle.
   (b) Let AD, BE, CF be cevian lines of triangle ABC. Prove that the radical center of circles constructed on AD, BE, CF as diameters is the orthocenter of the triangle.

7. If the common chord of two intersecting circles C_1 and C_2 is a diameter of C_2, circle C_2 is said to be *bisected* by circle C_1, and circle C_1 is said to *bisect* circle C_2. Prove the following theorems concerning bisected circles.
   (a) If circle C_2 is bisected by circle C_1 the square of the radius of C_2 is equal to the negative of the power of the center of C_2 with respect to C_1.
   (b) If point P lies inside a circle C_1, but not at the center of C_1, then P is the center of a circle C_2 that is bisected by C_1.
   (c) If the radical center of three circles with noncollinear centers lies inside the three circles, then it is the center of a circle which is bisected by each of the three circles.
   (d) The locus of the center of a circle which bisects two given nonconcentric circles is a straight line parallel to the radical axis of the two given circles. (This line is called the *antiradical axis* of the two given circles.)
   (e) The circles having their centers on a fixed line and bisecting a given circle form a coaxial pencil of circles.

8. Prove that if each of a pair of circles cuts each of a second pair orthogonally, then the radical axis of either pair is the line of centers of the other.

9. (a) Through a given point draw a circle that is orthogonal to two given circles.
   (b) Through a given point draw a circle that is coaxial with two given circles.

10. Construct a circle such that tangents to it from three given points shall have given lengths.

11. Prove *Casey's Power Theorem*: The difference of the powers of a point with

respect to two circles is equal to twice the product of the distance of the point from the radical axis and the distance between the centers of the circles.

12. (a) What is the locus of a point whose power with respect to a given circle is constant?
    (b) What is the locus of a point the sum of whose powers with respect to two circles is constant?
    (c) What is the locus of a point the difference of whose powers with respect to two circles is constant?
    (d) What is the locus of a point the ratio of whose powers with respect to two circles is constant?

13. (a) Prove that if a circle has its center on the radical axis of a coaxial pencil of circles and is orthogonal to one of the circles of the pencil, it is orthogonal to all the circles of the pencil.
    (b) Prove that if a circle is orthogonal to two circles of a coaxial pencil of circles, it is orthogonal to all the circles of the pencil.

14. Show that the radical axes of the circles of a coaxial pencil with a circle not belonging to the pencil are concurrent.

15. Given three lines, and on each a pair of points such that a circle passes through each two pairs, show that either the three lines are concurrent or the six points lie on one circle.

BIBLIOGRAPHY

ADLER, C. F., *Modern Geometry*, 2d ed. New York: McGraw-Hill, 1967.

ALTSHILLER-COURT, NATHAN, *College Geometry, an Introduction to the Modern Geometry of the Triangle and the Circle*. New York: Barnes and Noble, 1952.

______, *Modern Pure Solid Geometry*. New York: The Macmillan Company, 1935.

COOLIDGE, J. L., *A Treatise on the Circle and the Sphere*. New York: Oxford University Press, 1916.

COXETER, H. S. M., and S. L. GREITZER, *Geometry Revisited* (New Mathematical Library, No. 19). New York: Random House and The L. W. Singer Company, 1967.

DAUS, P. H., *College Geometry*. Englewood Cliffs, N.J.: Prentice-Hall, Inc., 1941.

DAVIS, D. R., *Modern College Geometry*. Reading, Mass.: Addison-Wesley, 1954.

DURRELL, C. V., *Modern Geometry*. New York: The Macmillan Company, 1928.

FORDER, H. G., *Higher Course Geometry*. New York: Cambridge University Press, 1949.

JOHNSON, R. A., *Modern Geometry, an Elementary Treatise on the Geometry of the Triangle and the Circle*. Boston: Houghton Mifflin Company, 1929.

KAY, D. C., *College Geometry*. New York: Holt, Rinehart and Winston, 1969.

LACHLAN, R., *An Elementary Treatise on Modern Pure Geometry*. New York: The Macmillan Company, 1893.

M'Clelland, W. J., *A Treatise on the Geometry of the Circle*. New York: The Macmillan Company, 1891.

Maxwell, E. A., *Geometry for Advanced Pupils*. New York: Oxford University Press, 1949.

Miller, L. H., *College Geometry*. New York: Appleton-Century-Crofts, 1957.

Narayan, Shanti, and M. L. Kochhar, *A Text Book of Modern Pure Geometry*. Delhi: S. Chand and Company (no date).

Pedoe, D., *Circles*. New York: Pergamon Press, 1957.

Perfect, Helen, *Topics in Geometry*. New York: The Macmillan Company, 1963.

Richardson, G., and A. S. Ramsey, *Modern Plane Geometry*. New York: The Macmillan Company, 1928.

Russell, J. W., *A Sequel to Elementary Geometry*. New York: Oxford University Press, 1907.

Shively, L. S., *An Introduction to Modern Geometry*. New York: John Wiley and Sons, 1939.

Taylor, E. H., and G. C. Bartoo, *An Introduction to College Geometry*. New York: The Macmillan Company, 1949.

Winsor, A. S., *Modern Higher Plane Geometry*. Boston: The Christopher Publishing House, 1941.

2

Elementary Transformations

One of the most useful methods exploited by geometers of the modern era is that of cleverly transforming a figure into another which is better suited to a geometrical investigation. The gist of the idea is this. We wish to solve a difficult problem connected with a given figure. We *transform* the given figure into another which is related to it in a definite way and such that under the transformation the difficult problem concerning the original figure becomes a simpler problem concerning the new figure. We *solve* the simpler problem related to the new figure, and then *invert* the transformation to obtain the solution of the more difficult problem related to the original figure.

The idea of solving a difficult problem by means of an appropriate transformation is not peculiar to geometry but is found throughout mathematics.* For example, if one were asked to find the Roman numeral representing the product of the two given Roman numerals LXIII and XXIV, one would *transform* the two given Roman numerals into the corresponding Hindu-Arabic numerals, 63 and 24, *solve* the related prob-

* See M. S. Klamkin and D. J. Newman, "The philosophy and applications of transform theory," *SIAM Review*, vol. 3, no. 1, Jan. 1961, pp. 10–36.

lem in the Hindu-Arabic notation by means of the familiar multiplication algorithm to obtain the product 1512, then *invert* this result back into Roman notation, finally obtaining MDXII as the answer to the original problem. By an appropriate transformation, a difficult problem has been converted into an easy problem.

Again, suppose we wish to show that the equation

$$x^7 - 2x^5 + 10x^2 - 1 = 0$$

has no root greater than 1. By the substitution $x = y + 1$ we *transform* the given equation into

$$y^7 + 7y^6 + 19y^5 + 25y^4 + 15y^3 + 11y^2 + 17y + 8 = 0.$$

Since the roots of this new equation are equal to the roots of the original equation diminished by 1 ($y = x - 1$), we must show that the new equation has no root greater than 0. We *solve* this problem simply by noting that all the coefficients in the new equation are positive, whence y cannot also be positive and yet yield a zero sum. Now if we *invert* the transformation we obtain the desired result.

Geometrical transformations, as indeed transformations in other areas of mathematics, are useful not only in solving problems, but also in discovering new facts. We *transform* a given figure into a new figure; by studying the new figure we *discover* some property in it; then we *invert* to obtain a property of the original figure. In this chapter we shall examine some elementary geometrical transformations that can frequently be used to simplify the solution of geometrical problems or to discover new geometrical facts. Applications of the *transform-solve-invert* and *transform-discover-invert* procedures will appear both in this chapter and, along with further transformations, in Chapter 3.

2.1 TRANSFORMATION THEORY

As an illustration of the principal concept to be introduced in this section, consider the set B of all books in some specific library and the set P of all positive integers. Let us associate with each book of the library the number of pages in the book. In this way we make correspond to each element of set B a unique element of set P, and we say that "the set B has been mapped into the set P." As another illustration, let N be the set of all names listed in some given telephone directory and let A be the set of twenty-six letters of the alphabet. Let us associate with each name in the directory the last letter of the surname, thus making correspond to each element of set N a unique element of set A. This correspondence defines "a mapping of set N into set A." These are examples of the following formal definition.

2.1.1 Definitions and notation. If A and B are two (not necessarily distinct) sets, then a *mapping* of set A into set B is a correspondence that associates with each element a of A a unique element b of B. We write $a \rightarrow b$, and call b the *image* (or *map*) of a under the mapping, and we say that element a has been *carried into* (or *mapped into*) element b by the mapping. If every element of B is the image of some element of A, then we say that set A has been mapped *onto* set B.

Thus if A is the set $\{1,2,3,4\}$ and B the set $\{a,b,c\}$, the associations

$$1 \rightarrow a,\ 2 \rightarrow b,\ 3 \rightarrow b,\ 4 \rightarrow a$$

define a mapping of set A into set B. This, however, is not a mapping of set A onto set B, since under the mapping element c of B is not the image of any element of set A. On the other hand, the mapping induced by the associations

$$1 \rightarrow a,\ 2 \rightarrow b,\ 3 \rightarrow b,\ 4 \rightarrow c$$

is a mapping of set A onto set B, for now every element of B is the image of some element of A.

A very important kind of mapping of a set A onto a set B is one in which distinct elements of set A have distinct images in set B. We assign a special name to such mappings.

2.1.2 Definition. A mapping of a set A onto a set B in which distinct elements of A have distinct images in B is called a *transformation* (or *one-to-one mapping*) of A onto B.

2.1.3 Definitions and notation. If, in Definition 2.1.2, A and B are the same set, then the mapping is a transformation of a set A onto itself. In this case there may be an element of A that corresponds to itself. Such an element is called an *invariant element* (or *double element*) of the transformation. A transformation of a set A onto itself in which every element is an invariant element is called the *identity transformation* on A, and will, when no ambiguity is involved, be denoted by I.

2.1.4 Definition and notation. It is clear that a transformation of set A onto set B defines a second transformation, of set B onto set A, wherein an element of B is carried into the element of A of which it was the image under the first transformation. This second transformation is called the *inverse* of the first transformation. If T represents a transformation of a set A onto a set B, then the inverse transformation will be denoted by T^{-1}.

Thus if, among the married couples of a certain city, we let A be the set of husbands and B the set of wives, then the mapping that associates with each man of set A his wife in set B is a transformation of set A onto set B. The inverse of this transformation is the mapping of B onto A in which each woman in set B is associated with her husband in set A.

We now introduce the notion of product of two transformations, and examine some properties of products.

2.1.5 DEFINITIONS AND NOTATION. Let T_1 be a transformation of set A onto set B and T_2 a transformation of set B onto set C. The performance of transformation T_1 followed by transformation T_2 induces a transformation T of set A onto set C, wherein an element a of A is associated with the element c of C which is the image under T_2 of the element b of B which is the image under T_1 of element a of A. Transformation T is called the *product*, T_2T_1, of transformations T_1 and T_2, taken in this order. If the product transformation T_2T_1 exists, we say that T_2 is *compatible* with T_1.

Note that in the product T_2T_1, transformation T_1 is to be performed first, then transformation T_2. That is, we perform the component transformations from *right to left*. This is purely a convention and we could, as some writers do, have agreed to write the product the other way about. We adopt the present convention because it better fits the algebraic treatment of transformations encountered in analytic geometry.

2.1.6 THEOREM. *If T_2 is compatible with T_1, it does not follow that T_1 is compatible with T_2. If, however, both T_1 and T_2 are transformations of a set A onto itself, then necessarily both T_2 is compatible with T_1 and T_1 is compatible with T_2.*

The reader can easily construct an example where T_2T_1 exists but T_1T_2 does not exist. The second part of the theorem is quite obvious.

2.1.7 DEFINITION. A transformation T of a set A onto itself is said to be *involutoric* if $T^2 \equiv TT = I$.

2.1.8 THEOREM. *A product of two compatible transformations, even if each is a transformation of a set A onto itself, is not necessarily commutative; that is, if T_1T_2 and T_2T_1 both exist, we do not necessarily have $T_1T_2 = T_2T_1$.*

Let A be the set of all points of a plane on which a rectangular coordinate framework has been superimposed. Let T_1 be the transformation of A onto itself which carries each point of A into a point one unit in the direction of the positive x axis, and let T_2 be the transformation of A onto itself which rotates each point of A counterclockwise about the origin through 90°. Under T_2T_1 the point (1,0) is carried into the point (0,2), whereas under T_1T_2 it is carried into the point (1,1). It follows that $T_1T_2 \neq T_2T_1$.

2.1.9 Theorem. *Multiplication of compatible transformations is associative; that is, if T_1, T_2, T_3 are transformations such that T_2 is compatible with T_1 and T_3 with T_2, then $T_3(T_2T_1) = (T_3T_2)T_1$.*

Both $T_3(T_2T_1)$ and $(T_3T_2)T_1$ denote the resultant transformation obtained by first performing T_1, then T_2, then T_3.

We leave to the reader the establishment of the following three theorems.

2.1.10 Theorem. *If T is a transformation of set A onto itself, then (1) $TI = IT = T$, (2) $TT^{-1} = T^{-1}T = I$.*

2.1.11 Theorem. *If T and S are transformations of a set A onto itself, and if $TS = I$, then $S = T^{-1}$.*

2.1.12 Theorem. *If transformation T_2 is compatible with transformation T_1, then $(T_2T_1)^{-1} = T_1^{-1}T_2^{-1}$.*

We conclude with a definition which is basic in deeper geometrical studies.

2.1.13 Definitions. A nonempty set of transformations of a set A onto itself is said to constitute a *transformation group* if the inverse of every transformation of the set is in the set and if the product of any two transformations of the set is in the set. If, in addition, the product of every two transformations of the set is commutative, then the transformation group is said to be *abelian* (or *commutative*).

PROBLEMS

1. If A represents the set of all integers, which of the following mappings of A into itself are mappings of A onto itself? Which are transformations of A onto itself?

   (a) $a \to a + 5$ (b) $a \to a + a^2$
   (c) $a \to a^5$ (d) $a \to 2a - 1$
   (e) $a \to 5 - a$ (f) $a \to a - 5$

2. If R represents the set of all real numbers, which of the following mappings of R into itself are mappings of R onto itself? Which are transformations of R onto itself?

   (a) $r \to 2r - 1$ (b) $r \to r^2$
   (c) $r \to r^3$ (d) $r \to 1 - r$
   (e) $r \to r + r^2$ (f) $r \to 5r$

3. If R represents the set of all real numbers, is the mapping indicated by the association $r \rightarrow r^3 - r$ a mapping of R onto itself? Is it a transformation of R onto itself?

4. (a) Generalize Definition 2.1.5 for the situation where T_1 is a mapping of set A *into* set B and T_2 is a mapping of set B *into* set C.
   (b) Let T_1 and T_2 be the mappings of the set N of natural numbers into itself indicated by the associations $n \rightarrow n^2$ and $n \rightarrow 2n + 3$ respectively. Find the associations for the mappings T_1T_2, T_2T_1, T_1^2, T_2^2, $(T_1T_2)T_1$, $T_1(T_2T_1)$.

5. Supply a proof for Theorem 2.1.6.

6. Establish Theorem 2.1.10.

7. Establish Theorem 2.1.11.

8. Establish Theorem 2.1.12.

9. If T is a transformation of set A onto set B, show that $(T^{-1})^{-1} = T$.

10. If T is an involutoric transformation, show that $T = T^{-1}$.

11. If T_1, T_2, T_3 are transformations of a set A onto itself, show that $(T_3T_2T_1)^{-1} = T_1^{-1}T_2^{-1}T_3^{-1}$.

12. DEFINITION. If T and S are two transformations of set A onto itself, the transformation STS^{-1} is called the *transform of T by S*.
    (a) Show that the transform of the inverse of T by S is the inverse of the transform of T by S.
    (b) If $TS = ST$, show that each transformation is its own transform by the other.
    (c) If T_1, T_2, S are all transformations of set A onto itself, show that the product of the transforms of T_1 and T_2 by S is the transform of T_2T_1 by S.

13. In abstract algebra a *group* is defined to be a nonempty set G of elements in which a binary operation $*$ is defined satisfying the following four postulates:
    G1: *For all a, b in G, $a * b$ is in G.*
    G2: *For all a, b, c in G, $(a * b) * c = a * (b * c)$.*
    G3: *There exists an element i of G such that, for all a in G, $a * i = a$.*
    G4: *For each element a of G there exists an element a^{-1} of G such that $a * a^{-1} = i$.*
    Show that a transformation group is a group in the sense of abstract algebra, where transformation multiplication plays the role of the binary operation.

2.2 FUNDAMENTAL POINT TRANSFORMATIONS OF THE PLANE

Let S be the set of all points of an ordinary plane. In this section we consider some fundamental transformations of the set S onto itself.

2.2.1 Definitions and notation. Let $\overline{AB}$ be a directed line segment in the plane. By the *translation* $T(AB)$ we mean the transformation of S onto itself which carries each point P of the plane into the point P' of the plane such that $\overline{PP'}$ is equal and parallel to $\overline{AB}$. The directed segment $\overline{AB}$ is called the *vector* of the translation.

2.2.2 Definitions and notation. Let O be a fixed point of the plane and θ a given sensed angle. By the *rotation* $R(O,\theta)$ we mean the transformation of S onto itself which carries each point P of the plane into the point P' of the plane such that $OP' = OP$ and $\measuredangle \overline{POP'} = \theta$. Point O is called the *center* of the rotation, and θ is called the *angle* of the rotation.

2.2.3 Definitions and notation. Let l be a fixed line of the plane. By the *reflection* $R(l)$ *in line* l we mean the transformation of S onto itself which carries each point P of the plane into the point P' of the plane such that l is the perpendicular bisector of PP'. The line l is called the *axis* of the reflection.

2.2.4 Definitions and notation. Let O be a fixed point of the plane. By the *reflection* (or *half-turn*) $R(O)$ *in* (*about*) *point* O we mean the transformation of S onto itself which carries each point P of the plane into the point P' of the plane such that O is the midpoint of PP'. Point O is called the *center* of the reflection.

2.2.5 Definitions and notation. Let O be a fixed point of the plane and k a given nonzero real number. By the *homothety* (or *expansion*, or *dilatation*, or *stretch*) $H(O,k)$ we mean the transformation of S onto itself which carries each point P of the plane into the point P' of the plane such that $\overline{OP'} = k\,\overline{OP}$. The point O is called the *center* of the homothety, and k is called the *ratio* of the homothety.

There are certain products of the above transformations which also are of fundamental importance.

2.2.6 Definitions and notation. Let l be a fixed line of the plane and $\overline{AB}$ a given directed segment on l. By the *glide-reflection* $G(l,AB)$ we mean the product $R(l)T(AB)$. The line l is called the *axis* of the glide-reflection, and the directed segment $\overline{AB}$ on l is called the vector of the glide-reflection.

2.2.7 Definitions and notation. Let l be a fixed line of the plane and O a fixed point on l, and let k be a given nonzero real number. By the *stretch-reflection* $S(O,k,l)$ we mean the product $R(l)H(O,k)$. The line l is called the *axis* of the stretch-reflection, the point O is called the *center* of the stretch-reflection, and k is called the *ratio* of the stretch-reflection.

2.2.8 DEFINITIONS AND NOTATION. Let O be a fixed point of the plane, k a given nonzero real number, and θ a given sensed angle. By the *homology** (*or stretch-rotation*, or *spiral rotation*) $H(O,k,\theta)$ we mean the product $R(O,\theta)H(O,k)$. Point O is called the *center* of the homology, k the *ratio* of the homology, and θ the *angle* of the homology.

The following theorems are easy consequences of the above definitions.

2.2.9 THEOREM. *If n is an integer, then $R(O,(2n + 1)180°) = R(O) = H(O, -1)$.*

2.2.10 THEOREM. *If n is an integer, then (1) $H(O,k,n360°) = H(O,k)$, (2) $H(O,k,(2n + 1)180°) = H(O, -k)$.*

2.2.11 THEOREM. $T(BC)T(AB) = T(AB)T(BC) = T(AC)$.

2.2.12 THEOREM. $R(O,\theta_1)R(O,\theta_2) = R(O,\theta_2)R(O,\theta_1) = R(O,\theta_1 + \theta_2)$.

2.2.13 THEOREM. $R(O,\theta)H(O,k) = H(O,k)R(O,\theta) = H(O,k,\theta)$.

2.2.14 THEOREM. *If $\overline{AB}$ is on l, then $R(l)T(AB) = T(AB)R(l) = G(l,AB)$.*

2.2.15 THEOREM. *If O is on l, then $R(l)H(O,k) = H(O,k)R(l) = S(O,k,l)$.*

2.2.16 THEOREM. (1) $[T(AB)]^{-1} = T(BA)$, (2) $[R(O,\theta)]^{-1} = R(O, -\theta)$, (3) $[R(l)]^{-1} = R(l)$, (4) $[R(O)]^{-1} = R(O)$, (5) $[H(O,k)]^{-1} = H(O,1/k)$, (6) $[G(l,AB)]^{-1} = G(l,BA)$, (7) $[H(O,k,\theta)]^{-1} = H(O,1/k,-\theta)$.

2.2.17 THEOREM. (1) $T(AA) = I$, (2) $R(O,n360°) = I$, *where n is any integer*, (3) $H(O,1) = I$.

2.2.18 THEOREM. *$R(l)$ and $R(O)$ are involutoric transformations.*

2.2.19 THEOREM. *In the unextended plane (1) a translation of nonzero vector has no invariant points, (2) a rotation of an angle which is not a multiple of 360° has only its center as an invariant point, (3) a reflection in a line has only the points of its axis as invariant points, (4) a reflection in a point has only its center as an invariant point, (5) a homothety of ratio different from 1 has only its center as an invariant point.*

* The term *homology* appears elsewhere in mathematics in a different sense.

PROBLEMS

1. Let O, P, M, N, referred to a rectangular cartesian coordinate system, be the points (0,0), (1,1), (1,0), (2,0) respectively, and let l denote the x axis. Find the coordinates of the point P' obtained from the point P by the following transformations: (a) $T(OM)$, (b) $R(O,90°)$, (c) $R(l)$, (d) $R(M)$, (e) $R(O)$, (f) $H(O,2)$, (g) $H(N,-2)$, (h) $H(M,\frac{1}{2})$, (i) $G(l,MN)$, (j) $S(O,2,l)$, (k) $H(O,2,90°)$, (l) $H(N,2,45°)$.

2. (a) If $O_1 \neq O_2$, are the rotations $R(O_1, \theta_1)$ and $R(O_2,\theta_2)$ commutative?
   (b) Are $R(O)$ and $R(l)$ commutative?
   (c) Are $R(O_1O_2)$ and $R(O_1,\theta)$ commutative?
   (d) Are $T(AB)$ and $R(l)$ commutative?
   (e) Are $T(AB)$ and $R(O)$ commutative?

3. If AB is carried into $A'B'$ by a rotation, locate the center of the rotation.

4. Let P map into P' under a glide-reflection. (a) Show that PP' is bisected by the axis of the glide-reflection. (b) Show that the square of the glide-reflection is a translation of twice the vector of the glide-reflection.

5. Let $ABCD$ be a square with center O. Show that $R(B,90°)\, R(C,90°) = R(O)$.

6. Let S be the square whose vertices are A:(1,1), B:(−1,1), C:(−1,−1), D:(1,−1), and let O be the origin. (a) Show that S is carried into itself under each of the transformations: $R(x \text{ axis})$, $R(y \text{ axis})$, $R(AC)$, $R(BD)$, $R(O,90°)$, $R(O)$, $R(O,270°)$, I. (b) Show that the transformations of part (a) form a transformation group.

7. Show that each of the following sets of point transformations in a plane constitutes an abelian transformation group: (a) all translations, (b) all concentric rotations, (c) all concentric homotheties, (d) all concentric homologies.

8. Show that $R(O_2)R(O_1) = T(2O_1O_2)$.

9. (a) Show that $T(AB)R(O)$ is a reflection in point O' such that $\overline{OO'}$ is equal and parallel to $(\overline{AB})/2$.
   (b) Show that $R(O)T(AB)$ is a reflection in point O' such that $\overline{O'O}$ is equal and parallel to $(\overline{AB})/2$.
   (c) Show that $T(OO')R(O) = R(M)$, where M is the midpoint of OO'.

10. (a) Show that $R(O_3)R(O_2)R(O_1)$ is a reflection in point O such that $\overline{OO_3}$ is equal and parallel to $\overline{O_1O_2}$.
    (b) Show that $R(O_3)R(O_2)R(O_1) = R(O_1)R(O_2)R(O_3)$.

11. Show that $R(OO')R(O) = R(O)R(OO') = R(l)$, where l is the line through O perpendicular to OO'.

12. Show that if $O \neq O'$, then $R(O',-\theta)R(O,\theta)$ is a translation.

13. Where is a point P if its image under the homothety $H(O_1,k_1)$ coincides with its image under the homothety $H(O_2, k_2)$, $k_1 \neq k_2$?

14. Let l_1 and l_2 be two lines intersecting in a point O and let θ be the angle from l_1 to l_2. Show that $G(l_2,CD)G(l_1,AB) = T(CD)R(O,2\theta)T(AB)$. In particular, if $\theta = 90°$, then $G(l_2,CD)G(l_1,AB) = T(CD)R(O)T(AB)$.

2.3 APPLICATIONS OF THE HOMOTHETY TRANSFORMATION

Before continuing our study of geometrical transformations, we pause to consider a few applications of the homothety transformation.

We first describe a linkage apparatus, known as a *pantograph*, which was invented about 1603 by the German astronomer Christolph Scheiner (ca. 1575–1650) for mechanically copying a figure on an enlarged or reduced scale. The instrument is made in a variety of forms and can be

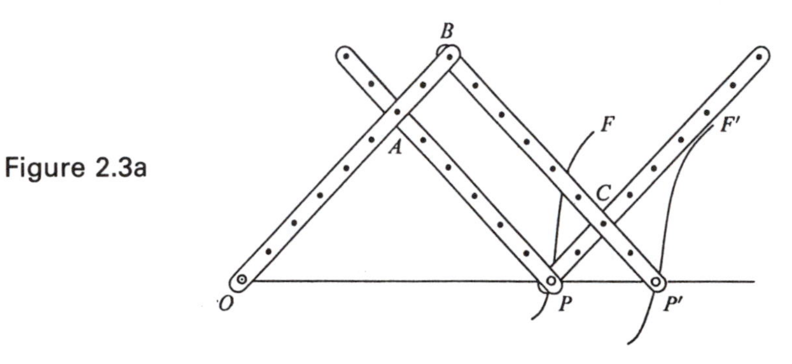

Figure 2.3a

purchased in a good stationery store. One form is pictured in Figure 2.3a, where the four equal rods are hinged by adjustable pivots at A, B, C, P, with $OA = AP$ and $PC = P'C = AB$. The instrument lies flat on the drawing paper and is fastened to the paper by a pointed pivot at O. Then if pencils are inserted at P and P', and P is made to trace a figure F, P' will trace the figure F' obtained from F by the homothety $H(O,\overline{OB}/\overline{OA})$. The reader can easily justify the working of the machine by showing that $APCB$ is a parallelogram, O, P, P' and collinear, and $\overline{OP'}/\overline{OP} = \overline{OB}/\overline{OA} =$ constant.

2.3.1 NOTATION. By the symbol $O(r)$ we mean the circle with center O and radius r.

2.3.2 DEFINITIONS. Let $A(a)$ and $B(b)$ be two nonconcentric circles and let I and E divide $\overline{AB}$ internally and externally in the ratio a/b. Then I

and E are called the *internal* and the *external centers of similitude* of the two circles (see Figure 2.3b).

Figure 2.3b

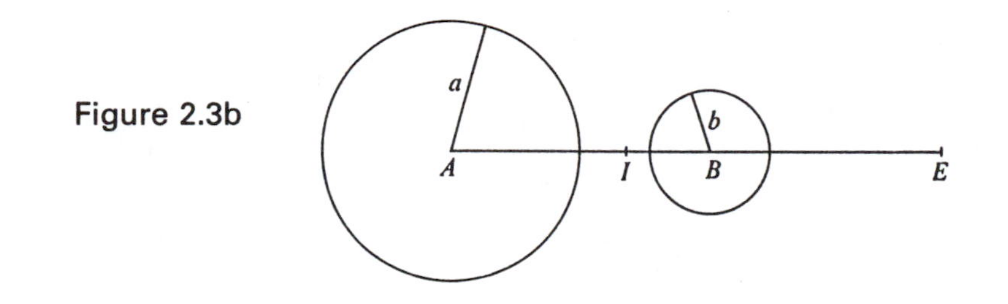

2.3.3 THEOREM. *Any two nonconcentric circles $A(a)$ and $B(b)$ with internal and external centers of similitude I and E are homothetic to each other under the homotheties $H(I,-b/a)$ and $H(E,b/a)$.*

Figure 2.3c

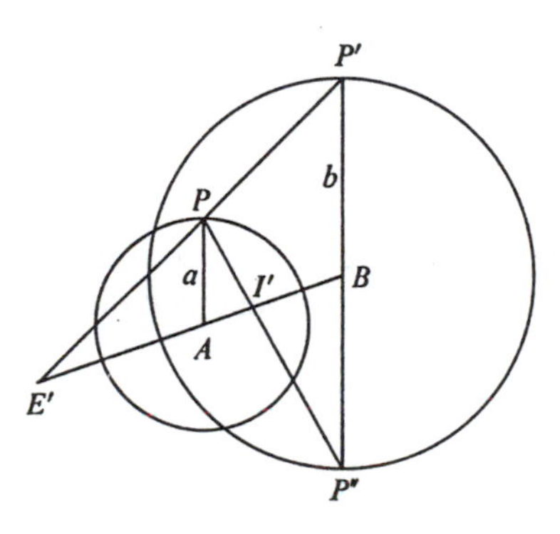

Let P (see Figure 2.3c) be any point on $A(a)$ not collinear with A and B. Let $P'BP''$ be the diameter of $B(b)$ parallel to AP, where $\overline{BP'}$ has the same direction as $\overline{AP}$. Let $P'P$ cut AB in E' and $P''P$ cut AB in I'. From similar triangles we find $\overline{E'P'}/\overline{E'P} = \overline{E'B}/\overline{E'A} = b/a$. Hence $E' = E$, the external center of similitude, and $B(b)$ is the image of $A(a)$ under the homothety $H(E,b/a)$. Similarly, $\overline{I'P''}/\overline{I'P} = \overline{I'B}/\overline{I'A} = -b/a$, and $I' = I$, the internal center of similitude. It follows that $B(b)$ is the image of $A(a)$ under the homothety $H(I,-b/a)$.

2.3.4 THEOREM. *The orthocenter H, the circumcenter O, and the centroid G of a triangle $A_1A_2A_3$ are collinear and $\overline{HG} = 2\overline{GO}$.*

Let M_1, M_2, M_3 (see Figure 2.3d) be the midpoints of the sides A_2A_3, A_3A_1, A_1A_2 of the triangle. Since $\overline{A_iG}/\overline{GM_i} = 2$ $(i = 1, 2, 3)$, triangle $M_1M_2M_3$ is carried into triangle $A_1A_2A_3$ by the homothety $H(G,-2)$. Therefore O, which is the orthocenter of triangle $M_1M_2M_3$, maps into the

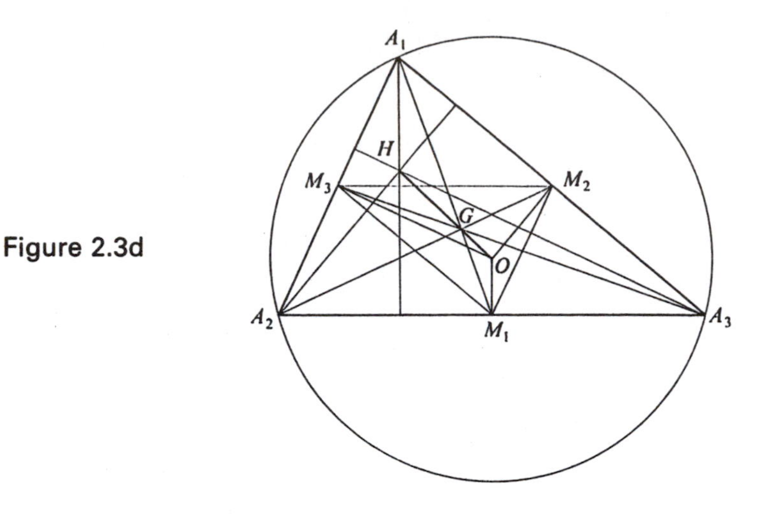

Figure 2.3d

orthocenter H of triangle $A_1A_2A_3$. It follows that H, G, O are collinear and $\overline{HG} = 2\,\overline{GO}$.

2.3.5 DEFINITION. The line of collinearity of the orthocenter, circumcenter, and centroid of a triangle is called the *Euler line* of the triangle.

2.3.6 THEOREM. *In triangle $A_1A_2A_3$ let M_1, M_2, M_3 be the midpoints of the sides A_2A_3, A_3A_1, A_1A_2, let H_1, H_2, H_3 be the feet of the altitudes on these sides, let N_1, N_2, N_3 be the midpoints of the segments A_1H, A_2H, A_3H, where H is the orthocenter of the triangle. Then the nine points M_1, M_2, M_3, H_1, H_2, H_3, N_1, N_2, N_3 lie on a circle whose center N is the midpoint of the segment joining the orthocenter H to the circumcenter O of the triangle, and whose radius is half the circumradius of the triangle.*

Referring to Figure 2.3e, we see that $\measuredangle A_2A_3H_3 = 90° - \measuredangle A_2 = \measuredangle A_2A_1S_1 = \measuredangle A_2A_3S_1$. Therefore right triangle HH_1A_3 is congruent to right triangle $S_1H_1A_3$, and H_1 is the midpoint of HS_1. Similarly, H_2 is the midpoint of HS_2 and H_3 is the midpoint of HS_3. Draw circumdiameter A_1T_1. Then T_1A_2 is parallel to A_3H_3 (since each is perpendicular to A_1A_2). Similarly, T_1A_3 is parallel to A_2H_2. Therefore $HA_3T_1A_2$ is a parallelogram and HT_1 and A_2A_3 bisect each other. That is, M_1 is the midpoint of HT_1. Similarly, M_2 is the midpoint of HT_2 and M_3 is the midpoint of HT_3. It now follows that the homothety $H(H,\frac{1}{2})$ carries A_1, A_2, A_3, S_1, S_2, S_3, T_1, T_2, T_3 into N_1, N_2, N_3, H_1, H_2, H_3, M_1, M_2, M_3, whence these latter nine points lie on a circle of radius half that of the circumcircle and with center N at the midpoint of HO.

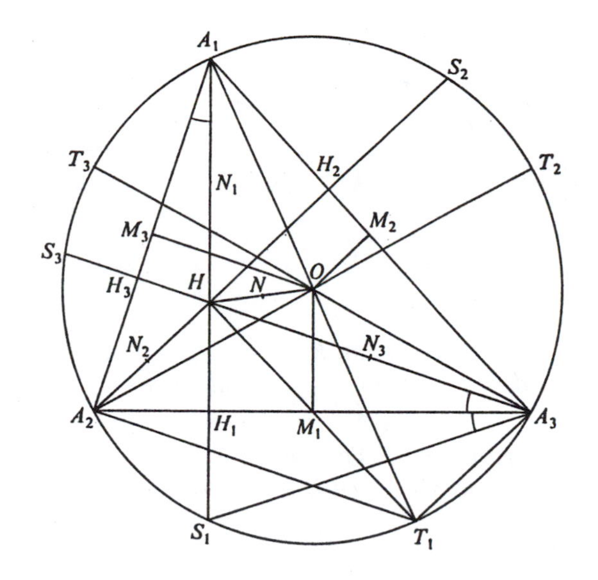

Figure 2.3e

2.3.7 Definition. The circle of Theorem 2.3.6 is called the *nine-point circle* of triangle $A_1A_2A_3$.

It was O. Terquem who named this circle the *nine-point circle*, and this is the name commonly used in the English-speaking countries. Some French geometers refer to it as *Euler's circle*, and German geometers usually call it *Feuerbach's circle*.

2.3.8 Definition. Let I and E be the internal and external centers of similitude of two given nonconcentric circles $A(a)$, $B(b)$ having unequal radii. Then the circle on IE as diameter is called the *circle of similitude* of the two given circles.

2.3.9 Theorem. *Let P be any point on the circle of similitude of two nonconcentric circles $A(a)$, $B(b)$ having unequal radii. Then $B(b)$ is the image of $A(a)$ under the homology $H(P, b/a, \measuredangle \overline{APB})$.*

Let I and E be the internal and external centers of similitude of the two given circles. If P coincides with I or E the theorem follows from Theorem 2.3.3. If P is distinct from I and E (see Figure 2.3f) then PI is perpendicular to PE and $(AB, IE) = -1$. Draw PA' so that PI bisects $\measuredangle A'PB$ internally. Then PE is the external bisector of the same angle, and it follows that $(A'B, IE) = -1$. Therefore $A' = A$ and $PB/PA = \overline{IB}/\overline{AI} = b/a$. The theorem now follows.

2.3.10 Corollary. *The locus of a point P moving in a plane such that the ratio of its distance from point A to its distance from point B of the*

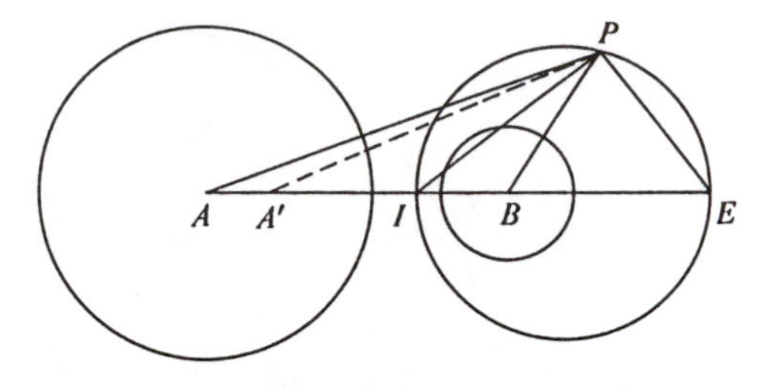

Figure 2.3f

plane is a positive constant $k \neq 1$ is the circle on IE as diameter, where I and E divide the segment AB internally and externally in the ratio k.

2.3.11 DEFINITION. The circle of Corollary 2.3.10 is called the *circle of Apollonius* of points A and B for the ratio k.

PROBLEMS

1. In Figure 2.3a show that O, P, P' are collinear and that $\overline{OP'}/\overline{OP} = \overline{OB}/\overline{OA}$.

2. In Figure 2.3g, AE, AB, BF represent three bars jointed at A and B. The

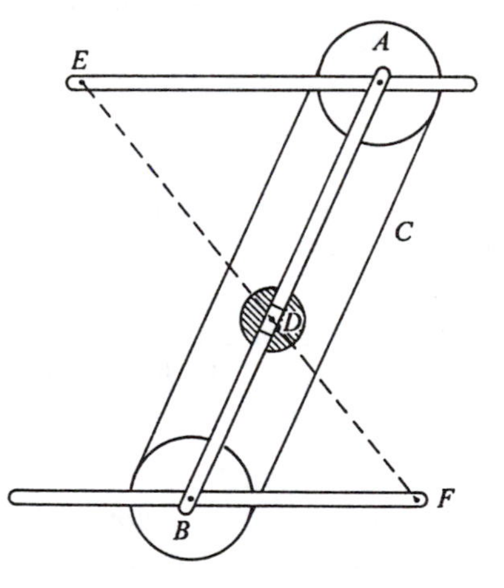

Figure 2.3g

bars AE and BF are attached at A and B to wheels of the same diameter, and around which goes a thin flexible steel band C. The result is that if bars AE and BF are so adjusted as to be parallel, they remain parallel however they are situated with respect to bar AB. The bars AE and BF are adjustable in length, and pencils are inserted at points E and F. D is a point adjustable along bar AB and about which the whole instrument can be rotated. Show that if D is fastened to AB so as to be collinear with E and F, then the pencil at E describes the map of a figure traced by the pencil at F under the homothety $H(D,-EA/BF)$.

3. Prove that if two circles have common external tangents, these tangents pass through the external center of similitude of the two circles, and if they have common internal tangents, these pass through the internal center of similitude of the two circles.

4. Let S be a center of similitude of two circles C_1 and C_2, and let one line through S cut C_1 in A and B and C_2 in A' and B', and a second line through S cut C_1 in C and D and C_2 in C' and D', where the primed points are the maps of the corresponding unprimed points under the homothety having center S and carrying circle C_1 into circle C_2. Show that: (a) $B'D'$ is parallel to BD, (b) A', C', D, B are concyclic and A, C, D', B' are concyclic, (c) $(SA')(SB) = (SA)(SB') = (SC')(SD) = (SC)(SD')$, (d) the tangents to C_1 and C_2 at B and A' intersect on the radical axis of C_1 and C_2.

5. Prove that the circle of similitude of two nonconcentric circles with unequal radii is the locus of points from which the two circles subtend equal angles.

6. Prove that two circles and their circle of similitude are coaxial.

7. (a) Show that the external centers of similitude of three circles with distinct centers taken in pairs are collinear.
   (b) Show that the external center of similitude of one pair of the circles and the internal centers of similitude of the other two pairs are collinear.

8. If a circle is tangent to each of two given nonconcentric circles, show that the line determined by the two points of tangency passes through a center of similitude of the two given circles.

9. If the distance between the centers of two circles $A(a)$ and $B(b)$ is c, locate the center of the circle of similitude of the two circles.

10. Show that any circle through the centers of two given nonconcentric circles of unequal radii is orthogonal to the circle of similitude of the two given circles.

11. In the notation of Theorems 2.3.4 and 2.3.6 show that:
    (a) $(HG,NO) = -1$.
    (b) The sum of the powers of the vertices A_1, A_2, A_3 with respect to the nine-point circle is $(A_2A_3^2 + A_3A_1^2 + A_1A_2^2)/4$.

12. Prove that the Circumcenter of the triangle formed by the tangents to the circumcircle of a given triangle at the vertices of the given triangle lies on the Euler line of the given triangle.

13. On the arc M_1H_1 of the nine-point circle, take the point X_1 one-third the way from M_1 to H_1. Take similar points X_2 and X_3 on arcs M_2H_2 and M_3H_3. Show that triangle $X_1X_2X_3$ is equilateral.

14. Show that if the Euler line is parallel to the side A_2A_3, then $\tan A_2 \tan A_3 = 3$, and $\tan A_2$, $\tan A_1$, $\tan A_3$ are in arithmetic progression.

15. Show that the trilinear polar (see Problem 11 Section 1.4) of the orthocenter of a triangle is perpendicular to the Euler line of the triangle.

2.4 ISOMETRIES

In this section and the next we consider those point transformations of the unextended plane that preserve all lengths and those that preserve all shapes. These are known, respectively, as *isometries* and *similarities*. We commence with a formal definition of these concepts.

2.4.1 DEFINITIONS. A point transformation of the unextended plane onto itself that carries each pair of points A, B into a pair A', B' such that $A'B' = k(AB)$, where k is a fixed positive number, is called a *similarity* (or an *equiform transformation*), and the particular case where $k = 1$ is called an *isometry* (or a *congruent transformation*). A similarity is said to be *direct* or *opposite* according as $\Delta\overline{ABC}$ has or has not the same sense as $\Delta\overline{A'B'C'}$. (A direct similarity is sometimes called a *similitude*, and an opposite similarity an *antisimilitude*. A direct isometry is sometimes called a *displacement*, and an opposite isometry a *reversal*.)

It is very interesting that isometries and similarities can be factored into products of certain of the fundamental point transformations considered in Section 2.2. We proceed to obtain some of these factorizations for the isometries.

2.4.2 THEOREM. *There is a unique isometry that carries a given noncollinear triad of points A, B, C into a given congruent triad A', B', C'.*

Superimposing the plane (by sliding, or turning it over and then sliding) upon its original position so that triangle ABC coincides with triangle $A'B'C'$ induces an isometry of the plane onto itself in which the point triad A, B, C is carried into the point triad A', B', C'. There is only the one isometry, for if P is any point in the plane, there is a unique point P' in the plane such that $P'A' = PA$, $P'B' = PB$, $P'C' = PC$.

2.4.3 THEOREM. *An isometry can be expressed as the product of at most three reflections in lines.*

Let an isometry carry the triad of points A, B, C into the congruent triad A', B', C'. We consider four cases. (1) If the two triads coincide, the isometry (by Theorem 2.4.2) is the identity I, which may be considered as the product of the reflection $R(l)$ with itself, where l is any line in the plane. (2) If A coincides with A' and B with B', but C and C' are distinct, the isometry (by Theorem 2.4.2) is the reflection $R(l)$, where l is the line AB. (3) If A coincides with A', but B and B' and C and C' are distinct, the reflection $R(l)$, where l is the perpendicular bisector of BB', reduces this case to one of the two previous cases. (4) Finally, if A and A', B and B', C and C' are distinct, the reflection $R(l)$, where l is the perpendicular bisector of AA', reduces this case to one of the first three cases. In each case, the isometry is ultimately expressed as a product of no more than three reflections in lines.*

2.4.4 THEOREM. *An isometry with an invariant point can be represented as the product of at most two reflections in lines.*

Let A be an invariant point of the isometry and let B and C be two points not collinear with A. Then the point triad A, B, C is carried into the triad A', B', C' where A' coincides with A. The desired result now follows from the first three cases in the proof of Theorem 2.4.3.

2.4.5 THEOREM. *Let l_1 and l_2 be any two lines of the plane intersecting in a point O, and let θ be the directed angle from l_1 to l_2, then $R(l_2)R(l_1) = R(O,2\theta)$. Conversely, a rotation $R(O,2\theta)$ can be factored into the product $R(l_2)R(l_1)$ of reflections in two lines l_1 and l_2 through O, where either line may be arbitrarily chosen through O and then the other such that the directed angle from l_1 to l_2 is equal to θ.*

Figure 2.4a

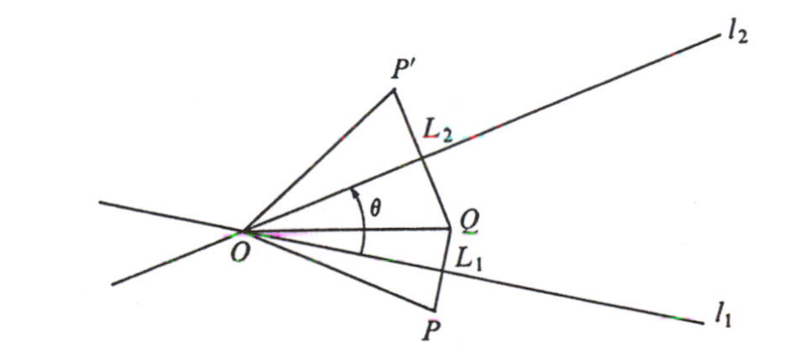

The proof is apparent from Figure 2.4a, since $OP' = OP$ and $\measuredangle \overline{POP'} = \measuredangle \overline{POQ} + \measuredangle \overline{QOP'} = 2 \measuredangle \overline{L_1OQ} + 2 \measuredangle \overline{QOL_2} = 2\theta$.

* One recalls a little test, which made the rounds of the mathematics meetings some years ago, for ferreting out incipient mathematicians. The test consists of two questions. (1) *You are in a room devoid of all furnishings except for a gas stove in one corner with one burner lit, and there is a kettle of water on the floor. What would you do to get the water in the kettle warm?* ANSWER: Place the kettle of water on the lighted burner. (2) *We have the same situation as in question* (1) *except that now there is also a table in the room and the kettle of water is on the table. What would you do to get the water in the kettle warm?* To give the same answer as before would be fatal. The correct answer is, "Remove the kettle from the table and place it on the floor, thus reducing the problem to one that has already been solved."

2.4.6 THEOREM. *Let l_1 and l_2 be any two parallel (or coincident) lines of the plane, and let $\overline{A_1A_2}$ be the directed distance from line l_1 to line l_2, then $R(l_2)R(l_1) = T(2A_1A_2)$. Conversely, a translation $T(2A_1A_2)$ can be factored into the product $R(l_2)R(l_1)$ of reflections in two lines l_1 and l_2 perpendicular to A_1A_2, where either line may be arbitrarily chosen perpendicular to A_1A_2 and then the other such that the directed distance from l_1 to l_2 is equal to $\overline{A_1A_2}$.*

Figure 2.4b

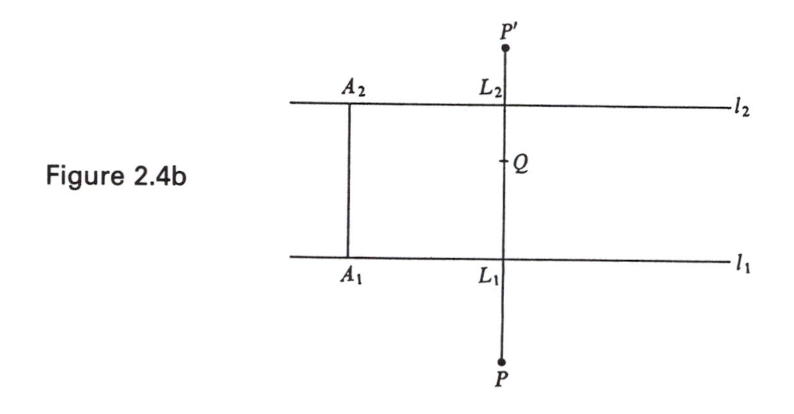

The proof is apparent from Figure 2.4b, since PP' is parallel to A_1A_2 and $\overline{PP'} = \overline{PQ} + \overline{QP'} = 2\overline{L_1Q} + 2\overline{QL_2} = 2\overline{A_1A_2}$.

2.4.7 THEOREM. *Any direct isometry is either a translation or a rotation.*

By Theorem 2.4.3, the isometry is a product of at most three reflections in lines. Since the isometry is direct, it must be a product of an *even* number of such reflections, and therefore of two such reflections. If the axes of the two reflections are parallel (or coincident), the isometry is a translation (by Theorem 2.4.6); otherwise the isometry is a rotation (by Theorem 2.4.5).

2.4.8 LEMMA. *$R(O)R(l) = G(m,2MO)$, where m is the line through O perpendicular to l and cutting l in point M.*

For (see Figure 2.4c), $\overline{PN} = \overline{QQ''} + \overline{Q''Q'} = 2(\overline{MQ''} + \overline{Q''O}) = 2\overline{MO}$ and $\overline{NQ'} = \overline{Q'P'}$.

2.4.9 THEOREM. *An opposite isometry T is either a reflection in a line or a glide-reflection.*

By Theorem 2.4.4, if T has an invariant point, T must be a reflection in a line. Suppose T has no invariant point and let T carry point A into point A'. Let O be the midpoint of AA'. Then $R(O)T$ is an opposite

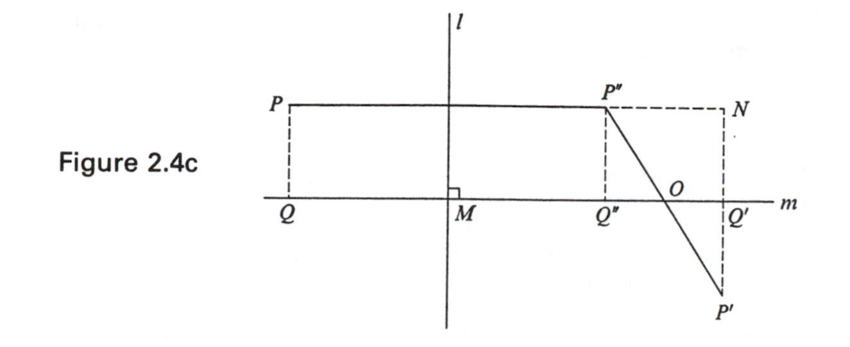

Figure 2.4c

isometry with invariant point A. Therefore $R(O)T = R(l)$, for some line l. That is, $T = [R(O)]^{-1}R(l) = R(O)R(l)$ = a glide-reflection, by Lemma 2.4.8.

2.4.10 THEOREM. *A product of three reflections in lines is either a reflection in a line or a glide-reflection.*

A product of three reflections in lines is an opposite isometry, and (by Theorem 2.4.9) an opposite isometry is either a reflection in a line or a glide-reflection.

PROBLEMS

1. Prove that an isometry maps straight lines into straight lines.

2. (a) Show that the product of two rotations is a rotation or a translation.
(b) Show that a direct isometry which is not a translation has exactly one invariant point.

3. (a) Prove that any isometry with an invariant point is a rotation or a reflection in a line according as it is direct or opposite.
(b) Prove that every opposite isometry with no invariant point is a glide-reflection.
(c) Prove that if an isometry has more than one invariant point, it must be either the identity or a reflection in a line.

4. (a) Show that $R(l)T(AB)$ is a glide-reflection whose axis is a line m parallel to l at a distance equal to one-half the projection of $\overline{BA}$ on a line perpendicular to l, and whose vector is the projection of $\overline{AB}$ on l.
(b) Show that $T(AB)R(l)$ is a glide-reflection whose axis is a line m parallel to l at a distance equal to one-half the projection of $\overline{AB}$ on a line perpendicular to l, and whose vector is the projection of $\overline{AB}$ on l.

5. Show that every opposite isometry is the product of a reflection in a line and a reflection in a point.

6. Show that $T(BA)R(O,\theta)T(AB) = R(O',\theta)$, where $\overline{O'O}$ is equal and parallel to $\overline{AB}$.

7. Show that $R(OO')R(O,\theta) = R(l)$, where l passes through O and the directed angle from l to OO' is $\theta/2$.

8. Let O, O' be two points on line l. Show that $G(l,2OO') = R(O')R(m)$, where m is the line through O perpendicular to l.

9. Let l_1, l_2, l_3 be the lines of the sides A_2A_3, A_3A_1, A_1A_2 of a triangle $A_1A_2A_3$.
   (a) Show that $G(l_3, A_1A_2)G(l_2, A_3A_1)G(l_1,A_2A_3)$ is a reflection.
   (b) Show that $[R(l_3)R(l_2)R(l_1)R(l_3)R(l_2)]^2$ is a translation along l_1.
   (c) Show that $G(l_3,A_1A_2)G(l_2,A_3A_1)G(l_1,A_2A_3) = R(l_3)R(A_2,2\sphericalangle A_3)$.
   (d) Show that $R(l_3)R(l_2)R(l_1) = R(l_3)R(A_3,2\sphericalangle A_3)$.
   (e) If triangle $A_1A_2A_3$ is acute, show that $R(l_3)R(l_2)R(l_1)$ is a glide-reflection with axis H_1H_3 and having direction $\overline{H_1H_3}$ and length equal to the perimeter of the orthic triangle $H_1H_2H_3$ of triangle $A_1A_2A_3$.

10. If l_1, l_2, l_3 are concurrent in a point O, show that $R(l_3)R(l_2)R(l_1)$ is a reflection in a line through O.

11. If l_1, l_2, l_3 are parallel, show that $R(l_3)R(l_2)R(l_1)$ is a reflection in a line parallel to l_1, l_2, l_3.

2.5 SIMILARITIES

We now examine the similarities.

2.5.1 THEOREM. *There is a unique similarity that carries a given noncollinear triad of points A, B, C into a given similar triad A', B', C'.*

If the triads are congruent, the unique similarity is the unique isometry guaranteed by Theorem 2.4.2. If the triads are not congruent, choose a point O of the plane. Now there is a homothety T_1 with center O carrying the triad A, B, C into a triad A'', B'', C'' congruent to the triad A', B', C', and (by Theorem 2.4.2) an isometry T_2 carrying triad A'', B'', C'' into the triad A', B', C'. Therefore the similarity T_2T_1 carries triad A, B, C into triad A', B', C'. But this is the only similarity, for if P is any point of the plane there is a unique point P' such that $P'A' = kPA$, $P'B' = kPB$, $P'C' = kPC$, where $k = A'B'/AB$.

2.5.2 THEOREM. *Every nonisometric similarity has a unique invariant point.*

Let $k \neq 1$ be the ratio of the similarity S and let S carry A_0 into A_1. There is no loss in generality in assuming $k < 1$, for (since $k \neq 1$) either

S or S^{-1} is actually a contraction, and S and S^{-1} have the same invariant points. If $A_1 = A_0$, then we have already found an invariant point. If $A_1 \neq A_0$, consider the sequence of points $A_0, A_1, A_2, A_3, \ldots$, where S carries A_i into A_{i+1}, $i = 0, 1, 2, \ldots$. If line segment A_0A_1 has length c, then A_1A_2 has length kc, A_2A_3 has length k^2c, etc. The circle of center A_0 and radius $c/(1-k)$ is carried into the circle of center A_1 and radius $kc/(1-k)$, then this circle into the circle of center A_2 and radius $k^2c/(1-k)$, etc.

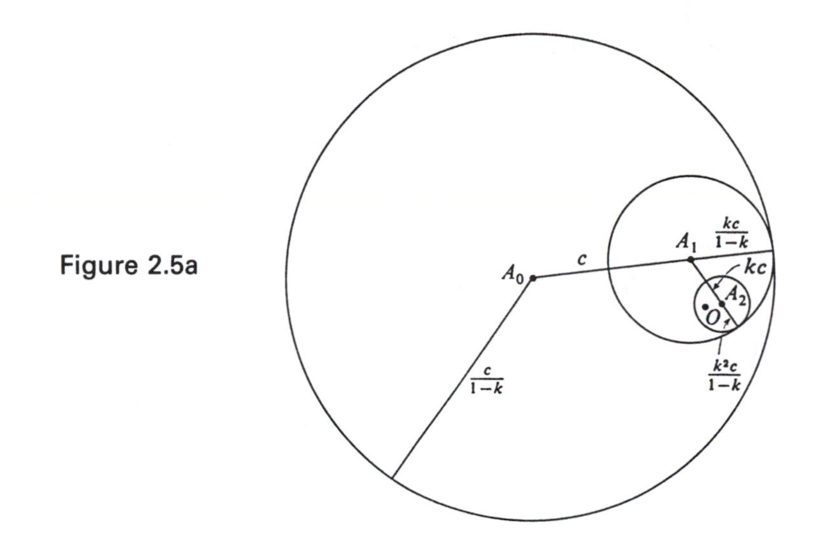

Figure 2.5a

Since $c + kc/(1-k) = c/(1-k)$, these circles (see Figure 2.5a) form a nested sequence of circles whose radii tend to zero as i increases. By the theorem of nested sets of analysis, the sequence of circles converges to a point of accumulation O. Since S carries $A_0, A_1, A_2, \ldots$ into $A_1, A_2, A_3, \ldots$, it follows that S leaves O invariant. Finally, S can have no more than one invariant point since the segment determined by two distinct invariant points would be mapped into itself, instead of being contracted by the ratio k.

2.5.3 THEOREM. *A direct similarity S is either a translation or a homology.*

If $k = 1$, S is a direct isometry and is then (by Theorem 2.4.7) either a translation or a rotation (which is a special homology). If $k \neq 1$, S has (by Theorem 2.5.2) an invariant point O. Then $S = TH(O,k)$, where T is a direct isometry with invariant point O. It follows that T is a rotation about O, and S is then a homology of center O.

2.5.4 THEOREM. *An opposite similarity S is either a glide-reflection or a stretch-reflection.*

If $k = 1$, S is an opposite isometry and is then (by Theorem 2.4.9) either a glide-reflection or a reflection in a line (which is a special glide-reflection). If $k \neq 1$, S has (by Theorem 2.5.2) an invariant point O. Then $S = TH(O,k)$, where T is an opposite isometry with invariant point O. It follows that T is a reflection in a line through O, and S is then a stretch-reflection of center O.

2.5.5 THEOREM. *A similarity that carries lines into parallel lines is either a translation or a homothety.*

The reader can easily show that this is a corollary of Theorems 2.5.3 and 2.5.4.

2.5.6 THEOREM. *If the line segments joining corresponding points of two given directly similar figures are divided proportionately, the locus of the dividing points is a figure directly similar to the given figures.*

By Theorem 2.5.3, the two given figures are related by a translation or a homology. The case of a translation presents no difficulty; the locus of the dividing points is clearly a figure directly congruent to each of the two given figures. Suppose, then, that the two given figures are related by a homology $H(O,k,\theta)$. Let A,A' be a fixed and P,P' a variable pair of

Figure 2.5b

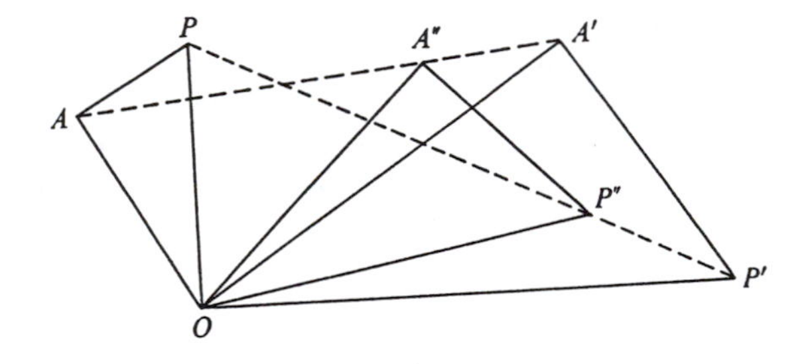

corresponding points in the two given figures (see Figure 2.5b). Then $OP'/OP = OA'/OA = k$ and $\sphericalangle \overline{POP'} = \sphericalangle \overline{AOA'} = \theta$. Let P'' and A'' be taken on AA' and PP' such that $\overline{PP''}/\overline{P''P'} = \overline{AA''}/\overline{A''A'}$. Since triangles POP' and AOA' are directly similar, it follows that triangles POP'' and AOA'' are also directly similar, and $OP''/OP = OA''/OA = k'$, say, and $\sphericalangle \overline{POP''} = \sphericalangle \overline{AOA''} = \theta'$, say. It follows that the locus of P'' is the image of the locus of P under the homology $H(O,k',\theta')$. That is, the locus of P'' is a figure directly similar to the two given figures.

PROBLEMS

1. Prove that a similarity maps straight lines into straight lines and circles into circles.

2. (a) Prove that any direct similarity which is not a translation has an invariant point.
(b) Prove that any opposite similarity which is not a glide-reflection has an invariant point.

3. (a) Show that if T is an opposite isometry, then T^2 is the identity or a translation.
(b) Show that if T is an opposite similarity, then T^2 is a translation or a homothety.

4. If l_1 is perpendicular to l_2 and O is their point of intersection, show that
(a) $S(O,k,l_1) = S(O,-k,l_2)$.
(b) lines l_1 and l_2 are invariant under $S(O,k,l_1)$.

5. Prove *Hjelmslev's Theorem*: When all the points P on one line are related by an isometry to all the points P' on another line, the midpoints of the segments PP' are distinct and collinear, or else they all coincide.

6. If two maps of the same country on different scales are drawn on tracing paper and then superposed, show that there is just one place that is represented by the same spot on both maps.

7. What is the product of (a) two stretch-reflections? (b) a homology and a stretch-reflection?

8. Give a proof of Theorem 2.5.5 utilizing Desargues' Theorem.

9. Prove the following theorem, which is Problem 4025 of *The American Mathematical Monthly* (Feb. 1943):

   Let $A'_1, A'_2, \ldots, A'_{2n}$ be the vertices of equilateral triangles constructed externally (or internally) on the sides A_1A_2, $A_2A_3, \ldots, A_{2n}A_1$ of a plane polygon of $2n$ sides $(P) = A_1A_2 \cdots A_{2n}$, and $M_1, M_2, \ldots, M_n$ be the midpoints of the principal diagonals A_1A_{n+1}, $A_2A_{n+2}, \ldots, A_nA_{2n}$ of (P). The midpoints $M'_1, M'_2, \ldots, M'_n$ of the principal diagonals $A'_2A'_{n+1}$, $A'_2A'_{n+2}, \ldots, A'_nA'_{2n}$ of the polygon $(P') = A'_1A'_2 \cdots A'_{2n}$ are the vertices of equilateral triangles constructed upon the sides of the polygon $(P) = M_1M_2 \cdots M_n$.

   Generalize by replacing the equilateral triangles by similar isosceles triangles.

10. (a) On the sides BC and CA of a triangle ABC, construct externally any two directly similar triangles, CBA_1 and ACB_1, Show that the midpoints of the three segments BC, A_1B_1, CA form a triangle directly similar to the two given triangles.
(b) On BC externally and on CA internally construct any two directly similar triangles CBA_1 and CAB_1. Show that the midpoints of AB and A_1B_1 form with C a triangle directly similar to the two given triangles.

These two problems constitute Problem E 521 of *The American Mathematical Monthly* (Jan. 1943).

2.6 INVERSION

In this section we briefly consider the inversion transformation, which is perhaps the most useful transformation we have for simplifying plane figures.

The history of the inversion transformation is complex and not clear-cut. Inversely related points were known to François Viète in the sixteenth century. Robert Simson, in his 1749 restoration of Apollonius' lost work *Plane Loci*, included (on the basis of commentary made by Pappus) one of the basic theorems of the theory of inversion, namely that the inverse of a straight line or a circle is a straight line or a circle. Simon A. J. L'Huilier (1750–1840) in his *Éléments d'analyse géométrique et d'analyse algébrique appliquées à la recherche des lieux géométriques* (Paris and Geneva, 1808) gave special cases of this theorem.

But inversion as a simplifying transformation for the study of figures is a product of more recent times, and was independently exploited by a number of writers. Bützberger has pointed out that Jacob Steiner disclosed, in an unpublished manuscript, a knowledge of the inversion transformation as early as 1824. It was refound in the following year by the Belgian astronomer and statistician Adolphe Quetelet. It was then found independently by L. I. Magnus, in a more general form, in 1831, by J. Bellavitis in 1836, then by J. W. Stubbs and J. R. Ingram, two Fellows of Trinity College, Dublin, in 1842 and 1843, and by Sir William Thomson (Lord Kelvin) in 1845. Thomson used inversion to give geometrical proofs of some difficult propositions in the mathematical theory of elasticity. In 1847 Liouville called inversion the *transformation by reciprocal radii*. Because of a property to be established shortly, inversion has also been called *reflection in a circle*.

2.6.1 DEFINITIONS AND NOTATION. If point P is not the center O of circle $O(r)$, the *inverse* of P in, or with respect to, circle $O(r)$ is the point P' lying on the line OP such that $(\overline{OP})(\overline{OP'}) = r^2$. Circle $O(r)$ is called the *circle of inversion*, point O the *center of inversion*, r the *radius of inversion*, and r^2 the *power of inversion*. We denote the inversion with center O and power $k > 0$ by the symbol $I(O,k)$.

From the above definition it follows that to each point P of the plane, other than O, there corresponds a unique inverse point P', and that if P' is the inverse of P, then P is the inverse of P'. Since there is no point corresponding, under the inversion, to the center O of inversion, we do not have a transformation of the set S of all points of the plane onto itself.

In order to make inversion a transformation, as defined in Definition 2.1.2, we may do either of two things. We may let S' denote the set of all points of the plane except for the single point O, and then inversion will be a transformation of the "punctured plane" S' onto itself. Or we may add to the set S of all points in the plane a single ideal "point at infinity" to serve as the correspondent under the inversion of the center O of inversion, and then the inversion will be a transformation of this augmented set S'' onto itself. It turns out that the second approach is the more convenient one, and we accordingly adopt the following convention.

2.6.2 CONVENTION AND DEFINITIONS. When working with inversion, we add to the set S of all points of the plane a single ideal point at infinity, to be considered as lying on every line of the plane, and this ideal point, Z, shall be the image under the inversion of the center O of inversion, and the center O of inversion shall be the image under the inversion of this ideal point Z.* The plane, augmented in this way, will be referred to as the *inversive plane*.

Of course, Convention 2.6.2 is at variance with the earlier Convention 1.2.1. But conventions are made only for convenience, and no trouble will arise if one states clearly which, if either, convention is being employed. For some investigations it is convenient to work in the ordinary plane, for others, in the extended plane, and for still others, in the inversive plane. It is to be understood that throughout the present section we shall be working in the inversive plane. The following theorem is apparent.

2.6.3 THEOREM. *Inversion is an involutoric transformation of the inversive plane onto itself which maps the interior of the circle of inversion onto the exterior of the circle of inversion and each point on the circle of inversion onto itself.*

One naturally wonders if there are any other self-inverse loci besides the circle of inversion. The next theorem deals with this matter. We recall that by "circle" is meant a straight line or a circle (see Definition 1.9.4).

2.6.4 THEOREM. *A "circle" orthogonal to the circle of inversion inverts into itself.*

This is obvious if the "circle" is a straight line, and the proof of Theorem 1.9.3 takes care of the case where the "circle" is a circle. Note that the "circle" inverts into itself as a whole and not point by point.

* When inverting the center O of inversion into the ideal point Z at infinity, one is reminded of Stephen Leacock's line about the rider who "flung himself upon his horse and rode off madly in all directions."

The following theorem suggests an easy way to construct the inverse of any given point distinct from the center of inversion.

2.6.5 THEOREM. *A point D outside the circle of inversion, and the point C where the chord of contact of the tangents from D to the circle of inversion cuts the diametral line OD, are inverse points.*

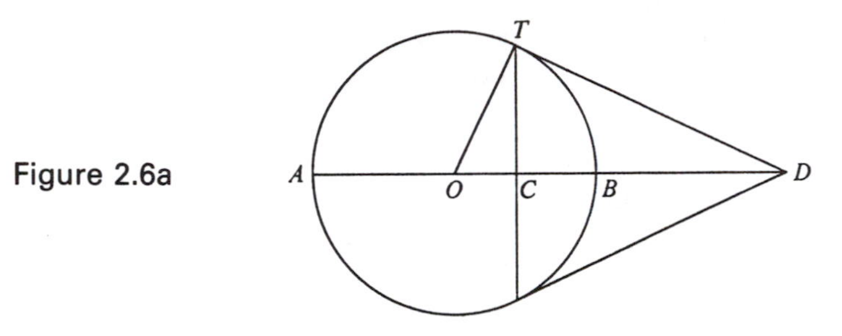

Figure 2.6a

For (see Figure 2.6a), $(\overline{OD})(\overline{OC}) = (OT)^2 = r^2$.

The following four theorems relate the concept of inverse points with some earlier concepts.

2.6.6 THEOREM. *If C, D are inverse points with respect to circle O(r), then (AB,CD) = −1, where AB is the diameter of O(r) through C and D; conversely, if (AB,CD) = −1, where AB is a diameter of circle O(r), then C and D are inverse points with respect to circle O(r).*

For (see Figure 2.6a), $(\overline{OC})(\overline{OD}) = r^2 = (OB)^2$ if and only if $(AB,CD) = -1$ (by Theorem 1.8.5).

2.6.7 THEOREM. *If C, D are inverse points with respect to circle O(r), then any circle through C and D cuts circle O(r) orthogonally; conversely, if a diameter of circle O(r) cuts a circle orthogonal to O(r) in C and D, then C and D are inverse points with respect to O(r).*

This, in view of Theorem 2.6.6, is merely an alternative statement of Theorem 1.9.3.

2.6.8 THEOREM. *If two intersecting circles are each orthogonal to a third circle, then the points of intersection of the two circles are inverse points with respect to the third circle.*

Let two circles intersect in points C and D and let O be the center of the third circle. Draw OC to cut the two given circles again in D' and D''. Then D' and D'' are each (by Theorem 2.6.7) the inverse of C with respect to the third circle. It follows that $D' = D'' = D$, and C and D are inverse points with respect to the third circle.

It is Theorem 2.6.8 that has led some geometers to refer to inversion as *reflection in a circle*. For if two intersecting circles are each orthogonal to a straight line, then the points of intersection of the two circles are reflections of each other in the line. Therefore, using the terminology "reflection in a circle" for "inversion with respect to a circle," and recalling Convention 2.6.2, we may subsume both the above fact and Theorem 2.6.8 in the single statement: *If two intersecting "circles" are each orthogonal to a third "circle," then the points of intersection of the two "circles" are reflections of each other in the third "circle."* Of course the same end can be achieved by using the terminology "inversion in a line" for "reflection in a line," and some geometers do just this.

2.6.9 THEOREM. $I(O,k_2)I(O,k_1) = H(O,k_2/k_1)$.

Let P be any point and let $I(O,k_1)$ carry P into P' and $I(O,k_2)$ carry P' into P''. Then, if $P \neq O$, we have O, P, P', P'' collinear and $(\overline{OP})(\overline{OP'}) = k_1$, $(\overline{OP'})(\overline{OP''}) = k_2$, whence O, P, P'' are collinear and $\overline{OP''}/\overline{OP} = k_2/k_1$. If $P = O$, then $P' = Z$, $P'' = O$. The theorem now follows.

When a point P traces a given curve C, the inverse point P' traces a curve C' called the *inverse* of the given curve. The next four theorems investigate the nature of the inverses of straight lines and circles. The first of the theorems has already been established as part of Theorem 2.6.4.

2.6.10 THEOREM. *The inverse of a straight line l passing through the center O of inversion is the line l itself.*

2.6.11 THEOREM. *The inverse of a straight line l not passing through the center O of inversion is a circle C passing through O and having its diameter through O perpendicular to l.*

Figure 2.6b

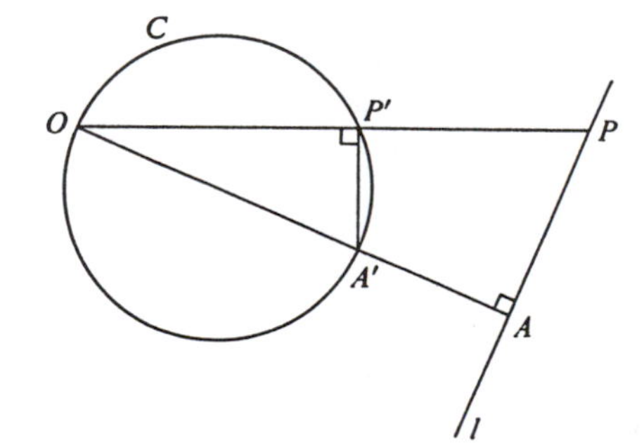

Let point A (see Figure 2.6b) be the foot of the perpendicular dropped from O on l. Let P be any other ordinary point on l and let A', P' be the inverses of A, P. Then $(\overline{OA})(\overline{OA'}) = (\overline{OP})(\overline{OP'})$, whence $OP'/OA' = OA/OP$ and triangles $OP'A'$, OAP are similar. Therefore $\sphericalangle OP'A' = \sphericalangle OAP = 90°$. It follows that P' lies on the circle C having OA' as diameter. Conversely, if P' is any point on circle C other than O or A', let OP' cut line l in P. Then, by the above, P' must be the inverse of P. Note that point O on circle C corresponds to the point Z at infinity on l.

2.6.12 THEOREM. *The inverse of a circle C passing through the center O of inversion is a straight line l not passing through O and perpendicular to the diameter of C through O.*

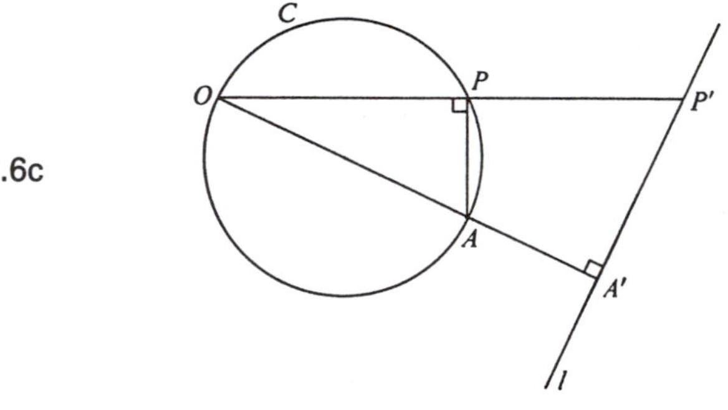

Figure 2.6c

Let point A (see Figure 2.6c) be the point on C diametrically opposite O, and let P be any point on circle C other than O and A. Let A', P' be the inverses of A, P. Then $(\overline{OA})(\overline{OA'}) = (\overline{OP})(\overline{OP'})$, whence $OP'/OA' = OA/OP$ and triangles $OP'A'$, OAP are similar. Therefore $\sphericalangle OA'P' = \sphericalangle OPA = 90°$. It follows that P' lies on the line l through A' and perpendicular to OA. Conversely, if P' is any ordinary point on line l other than A', let OP' cut circle C in P. Then, by the above, P' must be the inverse of P. Note that the point Z at infinity on line l corresponds to the point O on circle C.

2.6.13 THEOREM. *The inverse of a circle C not passing through the center O of inversion is a circle C′ not passing through O and homothetic to circle C with O as center of homothety.*

Let P (see Figure 2.6d) be any point on circle C. Let P' be the inverse of P and let OP cut circle C again in Q, Q coinciding with P if OP is tangent to circle C. Let r^2 be the power of inversion and let k be the power of point O with respect to circle C. Then $(\overline{OP})(\overline{OP'}) = r^2$ and $(\overline{OP})(\overline{OQ}) = k$, whence $\overline{OP'}/\overline{OQ} = r^2/k$, a constant. It follows that P' describes the map of the locus of Q under the homothety $H(O,r^2/k)$. That

Figure 2.6d

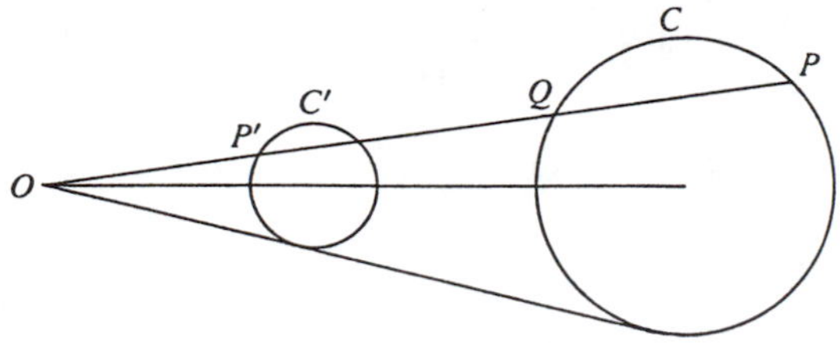

is, P' describes a circle C' homothetic to circle C and having O as center of homothety. Since circle C does not pass through O, circle C' also does not pass through O.

2.6.14 DEFINITION. A point transformation of a plane onto itself that carries "circles" into "circles" is called a *circular*, or *Möbius*, *transformation*.

Combining Theorems 2.6.10 through 2.6.13 we have

2.6.15 THEOREM. *Inversion is a circular transformation of the inversive plane.*

PROBLEMS

1. (a) Draw the figure obtained by inverting a square with respect to its center.
   (b) Draw the figure obtained by inverting a square with respect to one of its vertices.
2. (a) What is the inverse of a system of concurrent lines with respect to a point distinct from the point of concurrence?
   (b) What is the inverse of a system of parallel lines?
3. Prove that a coaxial system of circles inverts into a coaxial system of circles or into a set of concurrent or parallel lines.
4. (a) Let O be a point on a circle of center C, and let the inverse of this circle with respect to O as center of inversion intersect OC in B. If C' is the inverse of C, show that $OB = BC'$.
   (b) Show that the inverse C' of the center C of a given circle K is the inverse of the center O of inversion in the circle K' which is the inverse of the given circle K.
   (c) Calling reflection in a line inversion in the line, state the facts of parts (a) and (b) as a single theorem.
   (d) Prove that if two circles are orthogonal, the inverse of the center of either with respect to the other is the midpoint of their common chord.
5. Prove (for all cases): If two intersecting "circles" are each orthogonal to a third

"circle," then the points of intersection of the two "circles" are reflections of each other in the third "circle."

6. Show that isometries and similarities are circular transformations of the ordinary plane.

2.7 PROPERTIES OF INVERSION

We first establish a very useful theorem concerning directed angles between two "circles." We need the following lemma, whose easy proof will be left to the reader.

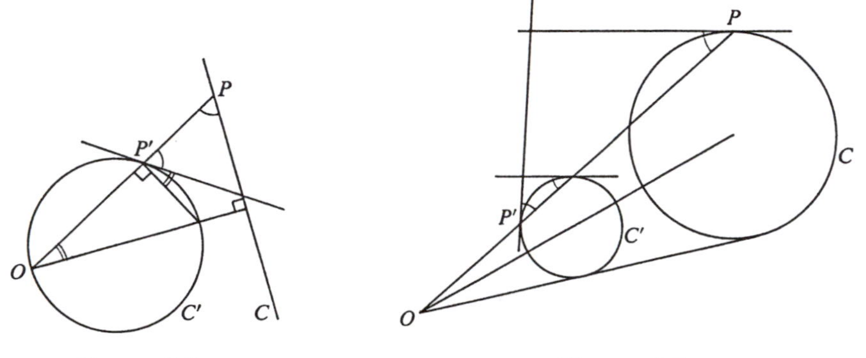

Figure 2.7a$_1$ Figure 2.7a$_2$

2.7.1 LEMMA. *Let C' (see Figure 2.7a$_1$, and 2.7a$_2$) be the inverse of "circle" C, and let P, P' be a pair of (perhaps coincident) corresponding points, under an inversion of center O, on C and C' respectively. Then the tangents (see Definition 1.9.4) to C and C' at P and P' are reflections of one another in the perpendicular to OP through the midpoint of PP'.*

2.7.2 THEOREM. *A directed angle of intersection of two "circles" is unaltered in magnitude but reversed in sense by an inversion.*

Let C and D be two "circles" intersecting in a point P, their inverses C' and D' intersecting in the inverse P' of P. Let c and d (see Figure 2.7b) be the tangents to C and D at P, and let c' and d' be the tangents to C' and D' at P'. Since, by Lemma 2.7.1, c and c', as well as d and d', are reflections of one another in the perpendicular to OP at the midpoint of PP', it follows that the directed angle from c to d is equal but opposite to the directed angle from c' to d'.

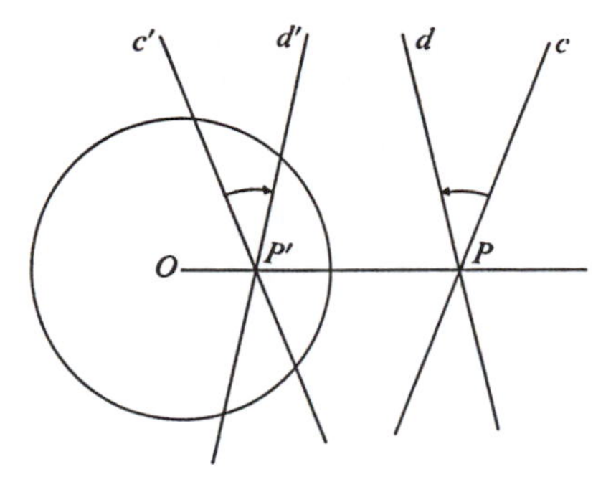

Figure 2.7b

In particular we have

2.7.3 COROLLARY. *(1) If two "circles" are tangent, their inverses are tangent. (2) If two "circles" are orthogonal, their inverses are orthogonal.*

There are other things, besides the magnitudes of angles between "circles," which remain invariant under an inversion transformation. Theorems 2.7.4 and 2.7.6 give two useful invariants of this sort. Theorem 2.7.5 is an important metrical theorem which shows how inversion affects distances between points. Theorem 2.7.7 is a sample of a whole class of theorems which are valuable when using inversion in its role of a simplifying transformation. When employing a particular transformation in geometry, it is of course important to know both the principal invariants of the transformation and some of the ways the transformation can simplify figures.

2.7.4 THEOREM. *(1) If a circle and two inverse points be inverted with respect to a center not on the circle, we obtain a circle and two inverse points. (2) If a circle and two inverse points be inverted with respect to a center on the circle, we obtain a straight line and two points which are reflections of one another in the straight line.*

Let points A and B (see Figure 2.7c) be inverse points with respect to a circle C and let K be any point. Draw circles C_1, C_2 through A and B but not through K. C_1, C_2 are orthogonal to C (by Theorem 2.6.7). Invert the figure with respect to center K.

(1) If K is not on C, we obtain circles C', C_1', C_2' and points A', B'. By Corollary 2.7.3, C_1' and C_2' are orthogonal to C', whence (by Theorem 2.6.8) A' and B' are inverse with respect to circle C'.

(2) If K is on C, we obtain a straight line C', circles C_1', C_2', and points A', B'. By Corollary 2.7.3, C_1' and C_2' are orthogonal to C'. It follows that A', B' are reflections of one another in line C'.

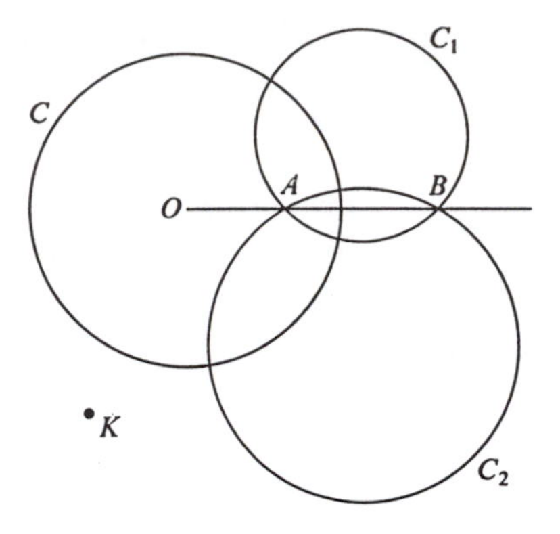

Figure 2.7c

2.7.5 Theorem. *If P, P' and Q, Q' are pairs of inverse points with respect to circle $O(r)$, then $P'Q' = (PQ)r^2/(OP)(OQ)$.*

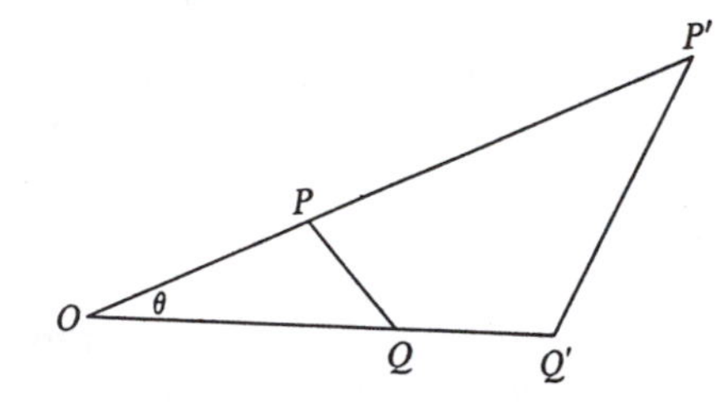

Figure 2.7d

Suppose (see Figure 2.7d) O, P, Q are not collinear. Since $(\overline{OP})(\overline{OP'}) = (\overline{OQ})(\overline{OQ'})$, triangle OPQ is similar to triangle $OQ'P'$, whence $P'Q'/PQ = OQ'/OP = (OQ')(OQ)/(OP)(OQ) = r^2/(OP)(OQ)$.

The case where O, P, Q are collinear follows from the above case by letting angle θ (see Figure 2.7d) approach zero. Or we may give a separate proof as follows (see Figure 2.7e):

Figure 2.7e

$$\begin{aligned}
(\overline{OP})(\overline{OP'}) &= (\overline{OQ})(\overline{OQ'}),\\
(\overline{OQ} + \overline{QP})\overline{OP'} &= \overline{OQ}(\overline{OP'} + \overline{P'Q'}),\\
(\overline{QP})(\overline{OP'}) &= (\overline{OQ})(\overline{P'Q'}),\\
\overline{P'Q'} = (\overline{QP})(\overline{OP'})/\overline{OQ} &= (\overline{QP})(\overline{OP'})(\overline{OP})/(\overline{OP})(\overline{OQ})\\
&= (\overline{QP})r^2/(\overline{OP})(\overline{OQ}).
\end{aligned}$$

2.7.6 Theorem. *The cross ratio of four distinct points on a "circle" is* *invariant under any inversion whose center is distinct from each of the four points.*

Let A, B, C, D be the four distinct points on a "circle" K, and let K invert into "circle" K'.

(1) If K and K' are both circles, then

$$\begin{aligned} (A'B',C'D') &= e(A'C'/C'B')/(A'D'/D'B') && \text{(by Theorem 1.6.7)} \\ &= e(AC/CB)/(AD/DB) && \text{(by Theorem 2.7.5)} \\ &= (AB,CD). && \text{(by Theorem 1.6.7)} \end{aligned}$$

(2) If K is a straight line and K' a circle, then

$$\begin{aligned} (A'B',C'D') &= e(A'C'/C'B')/(A'D'/D'B') && \text{(by Theorem 1.6.7)} \\ &= e(AC/CB)/(AD/DB) && \text{(by Theorem 2.7.5)} \\ &= (AB,CD). && \text{(see Problem 2, Section 1.5)} \end{aligned}$$

(3) If K is a circle and K' a straight line, then

$$\begin{aligned} (A'B',C'D') &= e(A'C'/C'B')/(A'D'/D'B') && \text{(see Problem 2, Section 1.5)} \\ &= e(AC/CB)/(AD/DB) && \text{(by Theorem 2.7.5)} \\ &= (AB,CD). && \text{(by Theorem 1.6.7)} \end{aligned}$$

(4) If K and K' are both straight lines, then

$$\begin{aligned} (A'B',C'D') &= e(A'C'/C'B')/(A'D'/D'B') && \text{(see Problem 2, Section 1.5)} \\ &= e(AC/CB)/(AD/DB) && \text{(by Theorem 2.7.5)} \\ &= (AB,CD). && \text{(see Problem 2, Section 1.5)} \end{aligned}$$

2.7.7 THEOREM. *Two nonintersecting circles can always be inverted into a pair of concentric circles.*

If the two circles are already concentric, the proof is trivial. Otherwise, let C_1 and C_2 be the pair of nonintersecting circles and let l be their radical axis. Using two points on l as centers, draw two circles D_1 and D_2 each orthogonal to both C_1 and C_2. Then (by Theorem 1.10.7 (1)) D_1 and D_2 intersect in two points, P_1, and P_2. Choose either of these two points, say P_1, as a center of inversion and invert the entire figure. C_1 and C_2 (by Theorem 2.6.13) become circles C_1' and C_2'. D_1 and D_2 (by Theorem 2.6.12) become straight lines D_1' and D_2', each of which (by Corollary 2.7.3 (2)) cuts circles C_1' and C_2' orthogonally. This means that C_1' and C_2' are concentric.

PROBLEMS

1. Two circles intersect orthogonally at P; O is any point on any circle touching the former circles at Q and R. Prove that the circles OPQ, OPR intersect at an angle of 45°.

2. If PQ, RS are common tangents to two circles PAR, QAS, prove that the circles PAQ, RAS are tangent to each other.

3. (a) Invert with respect to the center of the semicircle the theorem: An angle inscribed in a semicircle is a right angle.
(b) Invert with respect to A the theorem: If A, B, C, D are concyclic points, then angles ABD, ACD are equal or supplementary.

4. Prove that the circles having for diameters the three chords AB, AC, AD of a given circle intersect by pairs in three collinear points.

5. Given a triangle ABC and a point M, draw the circles MBC, MCA, MAB, and then draw the tangents to these circles at M to cut BC, CA, AB in R, S, T. Prove that R, S, T are collinear.

6. If A, B, C, D are four coplanar points with no three collinear, prove that circles ABC and ADC intersect at the same angle as the circles BDA and BCD.

7. Circles K_1, K_2 touch each other at T, and a variable circle through T cuts K_1, K_2 orthogonally in X_1, X_2 respectively. Prove that X_1X_2 passes through a fixed point.

8. A variable circle K touches a fixed circle K_1 and is orthogonal to another fixed circle K_2. Show that K touches another fixed circle coaxial with K_1 and K_2.

9. A, B, C, D are four concyclic points. If a circle through A and B touches one through C and D, prove that the locus of the point of contact is a circle.

10. What is the locus of the inverse of a given point in a system of tangent coaxial circles?

11. AC is a diameter of a given circle, and chords AB, CD intersect (produced if necessary) in a point O. Prove that circle OBD is orthogonal to the given circle.

12. Solve Problem 4, Section 2.6, parts (a) and (b), by means of Theorem 2.7.4.

2.8 APPLICATIONS OF INVERSION

We give a few illustrations of inversion as a simplifying transformation. We first emphasize the *transform-solve-invert* procedure, described in the introduction to this chapter, by an informal discussion of the problem (see Figure 2.8a$_1$):

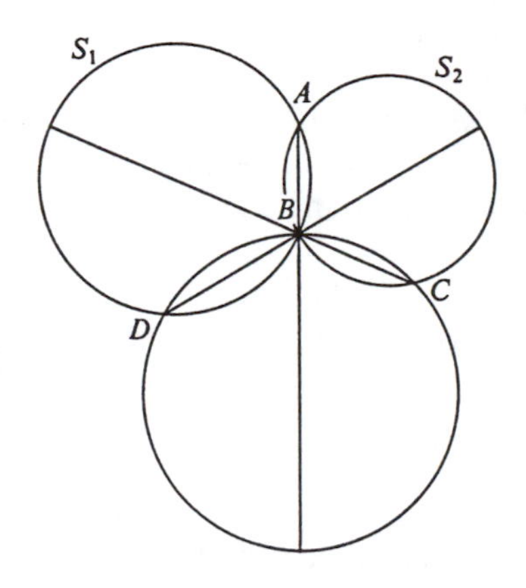

Figure 2.8a$_1$

Let two circles S_1 and S_2 intersect in A and B, and let the diameters of S_1 and S_2 through B cut S_2 and S_1 in C and D. Show that line AB passes through the center of circle BCD.

The figure of the problem involves three lines and three circles, all passing through a common point B. This suggests that we *transform* the figure into a simpler one by an inversion having center B, for under such an inversion the three lines will (by Theorem 2.6.10) map into themselves, and the three circles will (by Theorem 2.6.12) map into three lines. For convenience, we sketch the appearance of the simplified figure, not upon

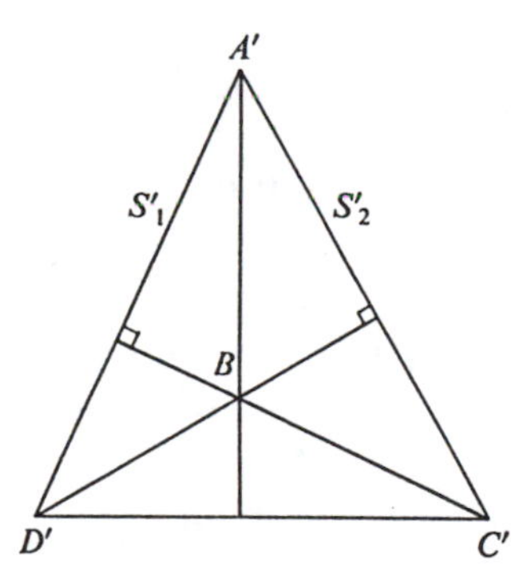

Figure 2.8a$_2$

the first figure, as it would naturally appear, but in Figure 2.8a$_2$. Note that circles BCD, $S_1 = ABD$, $S_2 = ABC$ have become straight lines $D'C'$, $A'D'$, $A'C'$, respectively, and the straight lines AB, CB, DB have become the straight lines $A'B$, $C'B$, $D'B$, respectively. Since line BC, being a diametral line of circle S_1, cuts S_1 orthogonally, we have (by Corollary

2.7.3 (2)) that BC' is perpendicular to $S_1' = A'D'$. Similarly, we have that BD' is perpendicular to $S_2' = A'C'$.

Now it is our desire to show that AB is a diametral line of circle BCD, or, in other words, to show that AB is orthogonal to circle BCD. We therefore attempt to *solve*, in the simplified figure, the allied problem: *Show that $A'B$ is perpendicular to $D'C'$*. But this is easily accomplished, for, since BC' and BD' are perpendicular to $A'D'$ and $A'C'$ respectively, B is the orthocenter of triangle $A'D'C'$, and $A'B$ must be perpendicular to $D'C'$.

Since $A'B$ is perpendicular to $D'C'$, if we *invert* the transformation that carried the first figure into the simplified one, we discover that AB is orthogonal to circle BCD, and our original problem is now solved.

The three-part procedure, *transform-solve-invert*, has carried us through. Our problem has turned out to be nothing but the inverse, with respect to the orthocenter as center of inversion, of the fact that the three altitudes of a triangle are concurrent, and the relation of the problem might well have been first discovered in just this way.

The five applications of inversion that now follow will be sketched only briefly, and the reader is invited to supply any missing details.

Ptolemy's Theorem

The following proposition was brilliantly employed by Claudius Ptolemy (85?–165?) for the development of a table of chords in the first book of his *Almagest*, the great definitive Greek work on astronomy. In all probability the proposition was known before Ptolemy's time, but his proof of it is the first that has come down to us. It is interesting that a very simple demonstration of the proposition—indeed, of an extension of the proposition—can be given by means of the inversion transformation.

2.8.1 PTOLEMY'S THEOREM. *In a cyclic convex quadrilateral the product of the diagonals is equal to the sum of the products of the two pairs of opposite sides.*

Referring to Figure 2.8b, subject the quadrilateral and its circumcircle to the inversion $I(A,1)$. The vertices B, C, D map into points B', C', D' lying on a straight line. It follows that $B'C' + C'D' = B'D'$, whence (by Theorem 2.7.5)

$$BC/(AB \cdot AC) + CD/(AC \cdot AD) = BD/(AB \cdot AD),$$

or

$$BC \cdot AD + CD \cdot AB = BD \cdot AC.$$

If quadrilateral $ABCD$ is not cyclic, then B', C', D' will not be collinear and $B'C' + C'D' > B'D'$. Using this fact the reader can easily supply a proof, fashioned after the above, for the following:

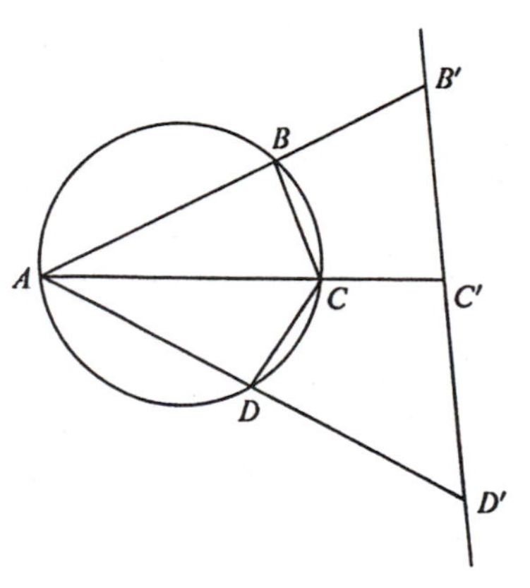

Figure 2.8b

2.8.2 EXTENSION OF PTOLEMY'S THEOREM. *In a convex quadrilateral $ABCD$,*

$$BC \cdot AD + CD \cdot AB \geqq BD \cdot AC,$$

with equality if and only if the quadrilateral is cyclic.

Pappus' Ancient Theorem

In Book IV of Pappus' *Collection* (ca. 300) appears the following beautiful proposition, referred to by Pappus as being already ancient in his time. The proof of the proposition by inversion is singularly attractive.

2.8.3 PAPPUS' ANCIENT THEOREM. *Let X, Y, Z be three collinear points with Y between X and Z, and let C, C_1, K_0 denote semicircles, all lying on the same side of XZ, on XZ, XY, YZ as diameters. Let K_1, K_2, $K_3, \ldots$ denote circles touching C and C_1, with K_1 also touching K_0, K_2 also touching K_1, K_3 also touching K_2, and so on. Denote the radius of K_n by r_n, and the distance of the center of K_n from XZ by h_n. Then $h_n = 2nr_n$.*

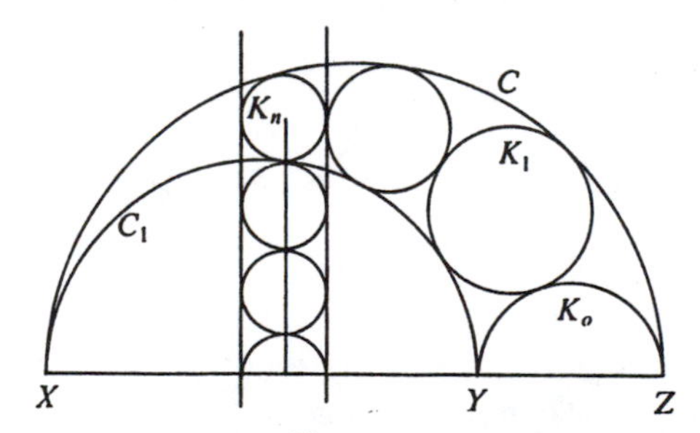

Figure 2.8c

Subject the figure to the inversion $I(X,t_n^2)$, where t_n is the tangent length from X to circle K_n. Then K_n inverts into itself (see Figure 2.8c), C and

C_1 invert into a pair of parallel lines tangent to K_n and perpendicular to XZ. K_0, K_1, K_2, . . .,K_{n-1} invert into a semicircle and circles, all of the same radius, tangent to the two parallel lines and such that K_1' touches K_0', K_2' touches K_1', . . . , K_{n-1}' touches K_{n-2}' and also K_n. It is now clear that $h_n = 2nr_n$.

Feuerbach's Theorem

Geometers universally regard the so-called Feuerbach's Theorem as undoubtedly one of the most beautiful theorems in the modern geometry of the triangle. The theorem was first stated and proved by Karl Wilhelm Feuerbach (1800–1834) in a work of his published in 1822; his proof was of a computational nature and employed trigonometry. A surprising number of proofs of the theorem have been given since, but probably none is as neat as the following proof employing the inversion transformation. We first state two definitions.

2.8.4 DEFINITION. A circle tangent to one side of a triangle and to the other two sides produced is called an *excircle* of the triangle. (There are four circles touching all three side lines of a triangle—the incircle and three excircles.)

2.8.5 DEFINITION. Two lines are said to be *antiparallel* relative to two transversals if the quadrilateral formed by the four lines is cyclic.

2.8.6 FEUERBACH'S THEOREM. *The nine-point circle of a triangle is tangent to the incircle and to each of the excircles of the triangle.*

Figure 2.8d

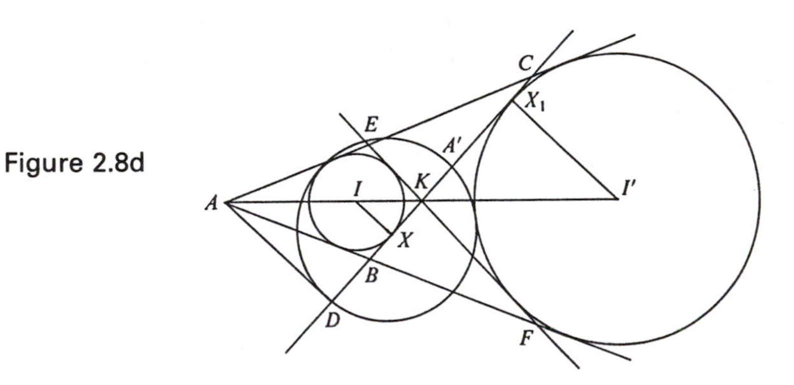

Figure 2.8d shows the incircle (I) and one excircle (I') of a triangle ABC. The four common tangents to these two circles determine the homothetic centers A and K lying on the line of centers II', and we have $(AK, II') = -1$. If D, X, X_1 are the feet of the perpendiculars from A, I, I'

on line BC, we then have $(DK,XX_1) = -1$. Now the line segments BC and XX_1 have a common midpoint A', and

$$(1) \qquad (A'K)(A'D) = (A'X)^2 = (A'X_1)^2.$$

Subject the figure to the inversion $I(A',\overline{A'X}^2)$. The circles (I) and (I') invert into themselves. Since the nine-point circle passes through A' and D, it follows that this circle inverts into a straight line through K, the inverse of D by (1). Also, the angle that this line makes with BC is equal but opposite to the angle that the tangent to the nine-point circle at D makes with BC, or therefore equal in both magnitude and sign to the angle that the tangent to the nine-point circle at A' makes with BC. But it is easily shown that this latter tangent is parallel to the opposite side of the orthic triangle, and therefore antiparallel to BC relative to AB and AC. But EF is antiparallel to BC relative to AB and AC. It follows that line EF is the inverse of the nine-point circle. Since this line is tangent to both (I) and (I'), we have that the nine-point circle is tangent to both (I) and (I'). That the nine-point circle is tangent to each of the other excircles can be shown in a like manner.

Steiner's Porism

Consider a circle C_1 lying entirely within another circle C_2, and a sequence of circles, $K_1, K_2, \ldots$, each having external contact with C_1 and internal contact with C_2, and such that K_2 touches K_1, K_3 touches K_2, and so on. A number of interesting questions suggest themselves in connection with such a figure. For example, can it ever be that the sequence $K_1, K_2, \ldots$ is finite, in the sense that finally a circle K_n of the sequence is reached which touches both K_{n-1} and K_1? Do the points of contact of the circles $K_1, K_2, \ldots$ lie on a circle? Do their centers lie on a circle? Etc. The figure was a dear one to Jacob Steiner, and he proved a number of remarkable properties of it. It seems that the best way to study the figure is by the inversion transformation. We content ourselves here with a proof of just one of Steiner's theorems. We first formulate a definition.

2.8.7 DEFINITION. A *Steiner chain of circles* is a sequence of circles, finite in number, each tangent to two fixed nonintersecting circles and to two other circles of the sequence.

2.8.8 STEINER'S PORISM. *If two given nonintersecting circles admit a Steiner chain, they admit an infinite number, all of which contain the same number of circles, and any circle tangent to the two given circles and surrounding either none or both of them is a member of such a chain.*

The proof is simple. By Theorem 2.7.7, the two given circles may be inverted into a pair of concentric circles, the circles of the Steiner chain then becoming a Steiner chain of equal circles for the two concentric circles. Since the circles of this associated Steiner chain may each be advanced cyclically in the ring in which they lie to form a similar chain, and this can be done in infinitely many ways, the theorem follows.

Theorem 2.8.8 is representative of a whole class of propositions in which there is a condition for a certain relation to subsist, but if the condition holds then the relation subsists infinitely often. Such propositions are called *porisms*. Three books of porisms by Euclid have been lost.

Peaucellier's Cell

An outstanding geometrical problem of the last half of the nineteenth century was to discover a linkage mechanism for drawing a straight line. A solution was finally found in 1864 by a French army officer, A. Peaucellier (1832–1913), and an announcement of the invention was made by A. Mannheim (1831–1906), a brother officer of engineers and inventor of the so-called Mannheim slide rule, at a meeting of the Paris Philomathic Society in 1867. But the announcement was little heeded until Lipkin, a young student of the celebrated Russian mathematician Chebyshev (1821–1894), independently reinvented the mechanism in 1871. Chebyshev had been trying to demonstrate the impossibility of such a mechanism. Lipkin received a substantial reward from the Russian Government, whereupon Peaucellier's merit was finally recognized and he was awarded the great mechanical prize of the Institut de France. Peaucellier's instrument contains 7 bars. In 1874 Harry Hart (1848–1920) discovered a 5-bar linkage for drawing straight lines, and no one has since been able to reduce this number of links or to prove that a further reduction is impossible. Both Peaucellier's and Hart's linkages are based upon the fact that the inverse of a circle through the center of inversion is a straight line.

The subject of linkages became quite fashionable among geometers, and many linkages were found for constructing special curves, such as conics, cardioids, lemniscates, and cissoids. In 1933, R. Kanayama published (in the *Tôhoku Mathematics Journal*, v. 37, 1933, pp. 294–319), a bibliography of 306 titles of papers and works on linkage mechanisms written between 1631 and 1931. It has been shown (in *Scripta Mathematica*, v. 2, 1934, pp. 293–294) that this list is far from complete, and of course many additional papers have appeared since 1931.

It has been proved that there exists a linkage for drawing any given algebraic curve, but that there cannot exist a linkage for drawing any transcendental curve. Linkages have been devised for mechanically solving algebraic equations.

2.8.9 PEAUCELLIER'S CELL. *In Figure 2.8e, let the points A and B of the jointed rhombus $PAP'B$ be joined to the fixed point O by means of equal bars OA and OB, $OA > PA$. Then, if all points of the figure are free to move except point O, the points P and P' will describe inverse curves*

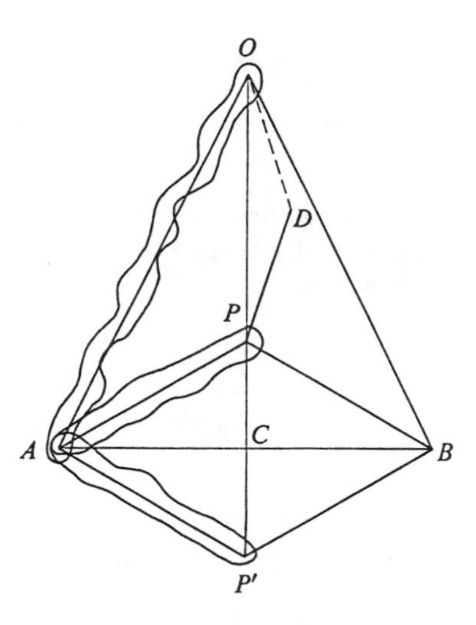

Figure 2.8e

under the inversion $I(O,OA^2 - PA^2)$. *In particular, if a seventh bar* DP, $DP > OP/2$, *is attached to* P *and the point* D *fixed so that* $DO = DP$, *then* P' *will describe a straight line.*

For we have

$$\begin{aligned}(OP)(OP') &= (OC - PC)(OC + PC)\\ &= OC^2 - PC^2\\ &= (OC^2 + CA^2) - (PC^2 + CA^2)\\ &= OA^2 - PA^2.\end{aligned}$$

PROBLEMS

1. Prove that if a circle C_1 inverts into a circle C_2, then the circle of similitude of C_1 and C_2 inverts into the radical axis of C_1 and C_2.

2. Circles OBC, OCA, OAB are cut in P, Q, R respectively by another circle through O. Prove that $(BP)(CQ)(AR) = (CP)(AQ)(BR)$.

3. Let $T_1T_2T_3T_4$ be a convex quadrilateral inscribed in a circle C. Let C_1, C_2, C_3, C_4 be four circles touching circle C externally at T_1, T_2, T_3, T_4 respectively. Show that

$$t_{12}t_{34} + t_{23}t_{41} = t_{13}t_{24},$$

where t_{ij} is the length of a common external tangent to circles C_i and C_j. (This is a special case of a more general theorem due to Casey. It can be considered as a generalization of Ptolemy's Theorem.)

4. (a) If $A(a)$ and $B(b)$ are two orthogonal circles, show that $I(B,b^2)I(A,a^2) = I(A,a^2)I(B,b^2)$.
(b) If K_1 K_2 are two orthogonal circles, A, A' inverse points in K_1, B and B' the inverses of A and A' in K_2, show that B, B' are inverse points in K_1.

5. If A, B, C, D are four concyclic points in the order A, C, B, D, and if p, q, r are the lengths of the perpendiculars from D to the lines AB, BC, CA respectively, show that

$$AB/p = BC/q + CA/r.$$

6. Let C' be the inverse of circle C under the inversion $I(O,r^2)$, and let p and p' be the powers of O with respect to C and C' respectively. Show that $pp' = r^4$.

7. Prove that the product of three inversions in three circles of a coaxial system is an inversion in a circle of that system.

8. We call the product $R(O)I(O,r^2)$ an *antinversion* in circle $O(r)$.
(a) Show that an antinversion is a circular transformation of the inversive plane.
(b) Show that a circle through a pair of antinverse points for a circle K cuts K diametrically.

9. Show that if two point triads inscribed in the same circle are copolar at a point C, then the inverse with respect to C of either triad is homothetic to the other triad.

10. (a) Show that two circles can be inverted into themselves from any point on their radical axis and outside both circles.
(b) When can three circles be inverted into themselves?

11. Show that a nonintersecting coaxial system of circles can be inverted into a system of concentric circles.

12. Show that any three circles can be inverted into three circles whose centers are collinear.

13. Show that any three points can, in general, be inverted into the vertices of a triangle similar to a given triangle.

14. Show that any three noncollinear points can be inverted into the vertices of an equilateral triangle of given size.

15. Circle C_1 inverts into circle C_2 with respect to circle C. Show that C_1 and C_2 invert into equal circles with respect to any point on circle C.

16. If a quadrilateral with sides a, b, c, x is inscribed in a semicircle of diameter x, show that

$$x^3 - (a^2 + b^2 + c^2)x - 2abc = 0.$$

(This is Problem E 574 of *The American Mathematical Monthly*, Feb. 1944.)

17. Prove Theorem 2.8.2.

18. Prove *Ptolemy's Second Theorem*: If $ABCD$ is a convex quadrilateral inscribed in a circle, then

$$AC/BD = (AB \cdot AD + CB \cdot CD)/(BA \cdot BC + DA \cdot DC).$$

19. (a) If A', B' are the inverses of A, B, then show that AA', BB' are antiparallel relative to AB and $A'B'$.
(b) If A, B, C, D are four points such that AB and CD are antiparallel relative to AD and BC, show that the four points can be inverted into the vertices of a rectangle.

20. Fill in the details of the proof of Theorem 2.8.6.

21. Show that the points of contact of the circles of a Steiner chain of circles all lie on a circle.

22. A linkage called *Hart's contraparallelogram* (invented by H. Hart in 1874) is

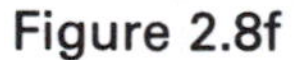
Figure 2.8f

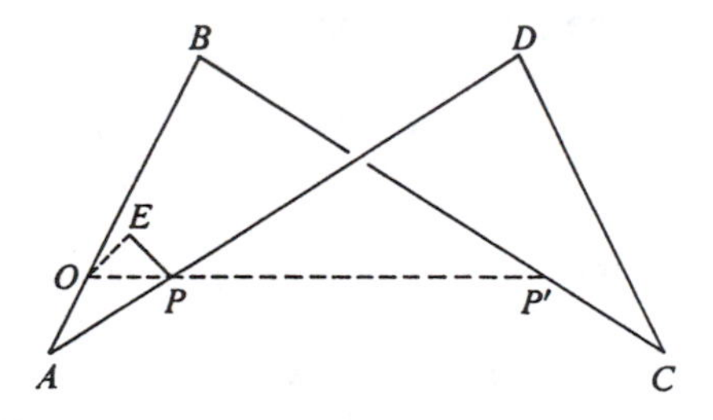

pictured in Figure 2.8f. The four rods AB, CD, BC, DA, $AB = CD$, $BC = DA$, are hinged at A, B, C, D. If O, P, P' divide AB, AD, CB proportionately, and the linkage is pivoted at O, show that O, P, P' are always collinear and that P and P' describe inverse curves with respect to O as center of the inversion. Hence if a fifth rod $EP > OP/2$ be pivoted at E with $OE = EP$, P' will describe a straight line.

23. Let four lines through a point V cut a circle in A, A'; B, B'; C, C'; D, D' respectively. Show that $(AB,CD) = (A'B',C'D')$. (This is Problem 13, Section 1.7.)

2.9 RECIPROCATION

We now consider a remarkable transformation of the set S of all points of the extended plane onto the set T of all straight lines of the extended plane.

2.9.1 DEFINITIONS AND NOTATION. Let $O(r)$ be a fixed circle (see Figure

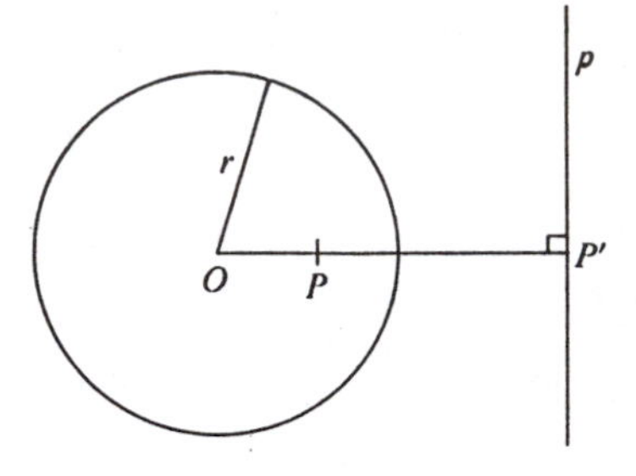

Figure 2.9a

2.9a) and let P be any ordinary point other than O. Let P' be the inverse of P in circle $O(r)$. Then the line p through P' and perpendicular to OPP' is called the *polar of* P for the circle $O(r)$. The *polar of* O is taken as the line at infinity, and the *polar of an ideal point* P is taken as the line through O perpendicular to the direction OP.

If line p is the polar of point P, then point P is called the *pole* of line p.

The pole-polar transformation set up by circle $K = O(r)$ will be denoted by $P(K)$ or $P(O(r))$ and will be called *reciprocation* in circle K.

Some nascent properties of reciprocation may be found in the works of Apollonius and Pappus. The theory was considerably developed by Desargues in his treatise on conic sections of 1639, and by his student Philippe de La Hire (1640–1718), and then greatly elaborated in the first half of the nineteenth century in connection with the study of the conic sections in projective geometry. The term *pole* was introduced in 1810 by the French mathematician F. J. Servois, and the corresponding term *polar* by Gergonne two to three years later. Gergonne and Poncelet developed the idea of poles and polars into a regular method out of which grew the elegant *principle of duality* of projective geometry.

The easy proof of the following theorem is left to the reader.

2.9.2 THEOREM. (*1*) *The polar of a point for a circle intersects the circle, is tangent to the circle at the point, or does not intersect the circle, according as the point is outside, on, or inside the circle.* (2) *If point P is outside a circle, then its polar for the circle passes through the points of contact of the tangents to the circle from P.*

The next theorem is basic in applications of reciprocation.

2.9.3 THEOREM. (*1*) *If, for a given circle, the polar of P passes through Q, then the polar of Q passes through P.* (2) *If, for a given circle, the pole of line p lies on line q, then the pole of q lies on p.* (3) *If, for a given* *circle, P and Q are the poles of p and q, then the pole of line PQ is the point of intersection of p and q.*

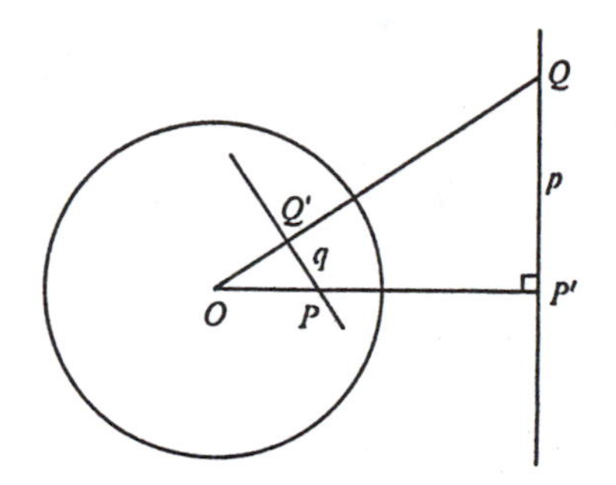

Figure 2.9b

(1) Suppose P and Q are ordinary points. Let P' (see Figure 2.9b) be the inverse of P and Q' the inverse of Q in the given circle, and suppose P' and Q are distinct. Then $\overline{OP} \cdot \overline{OP'} = \overline{OQ} \cdot \overline{OQ'}$, whence P, P', Q, Q' are concyclic and $\sphericalangle PP'Q = \sphericalangle PQ'O$. But, since Q lies on the polar of P, $\sphericalangle PP'Q = 90°$. Therefore $\sphericalangle PQ'O = 90°$, and P lies on the polar of Q. If $P' = Q$, the theorem is obvious. The cases where P, or Q, or both P and Q are ideal points are easily handled.

(2) Let P and Q be the poles of p and q. It is given that q (the polar of Q) passes through P. It follows, by (1), that p (the polar of P) passes through Q.

(3) Let p and q intersect in R. Then the polar of P passes through R, whence the polar of R passes through P. Similarly, the polar of R passes through Q. Therefore line PQ is the polar of R.

2.9.4 COROLLARY. *The polars, for a given circle, of a range of points constitute a pencil of lines; the poles, for a given circle, of a pencil of lines constitute a range of points.*

2.9.5 DEFINITIONS. Two points such that each lies on the polar of the other, for a given circle, are called *conjugate points* for the circle; two lines such that each passes through the pole of the other, for a given circle, are called *conjugate lines* for the circle.

The reader should find no difficulty in establishing the following facts.

2.9.6 THEOREM. *For a given circle: (1) Each point of a line has a conjugate point on that line. (2) Each line through a point has a conjugate line through that point. (3) Of two distinct conjugate points on a line that cuts the circle, one is inside and the other outside the circle. (4) Of two distinct conjugate lines that intersect outside the circle, one cuts the circle and the other does not. (5) Any point on the circle is conjugate to all the* *points on the tangent to the circle at the point.* (6) *Any tangent to the circle is conjugate to all lines through its point of contact with the circle.*

The next few theorems are important in the projective theory of poles and polars.

2.9.7 THEOREM. *If, for a given circle, two conjugate points lie on a line which intersects the circle, they are harmonically separated by the points of intersection.*

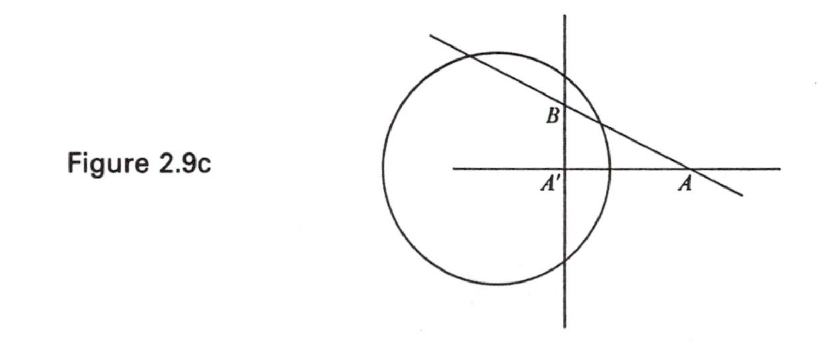

Figure 2.9c

Let A and B (see Figure 2.9c) be two such points, and let A' be the inverse of A in the circle. If $B = A'$, the desired result follows immediately. If $B \neq A'$, then $A'B$ is the polar of A and $\measuredangle AA'B = 90°$. The circle on AB as diameter, since it passes through A', is (by Theorem 2.6.7) orthogonal to the given circle. It follows (by Theorem 1.9.3) that A and B are harmonically separated by the points in which their line intersects the circle.

2.9.8 COROLLARY. *If a variable line through a given point intersects a circle, the harmonic conjugates of the point with respect to the intersections of the line and circle all lie on the polar of the given point.*

2.9.9 THEOREM. *If, for a given circle, two conjugate lines intersect outside the circle, they are harmonically separated by the tangents to the circle from their point of intersection.*

Let a and b (see Figure 2.9d) be two such lines and let S be their point of intersection. Since a and b are conjugate lines for the circle, the pole A of a lies on b, and the pole B of b lies on a. Then (by Theorem 2.9.3(3)) line AB is the polar of S and must pass through the points P and Q where the tangents from S touch the circle. Now A and B are conjugate points, whence (by Theorem 2.9.7) $(AB,PQ) = -1$. It follows that $(ba,pq) = -1$.

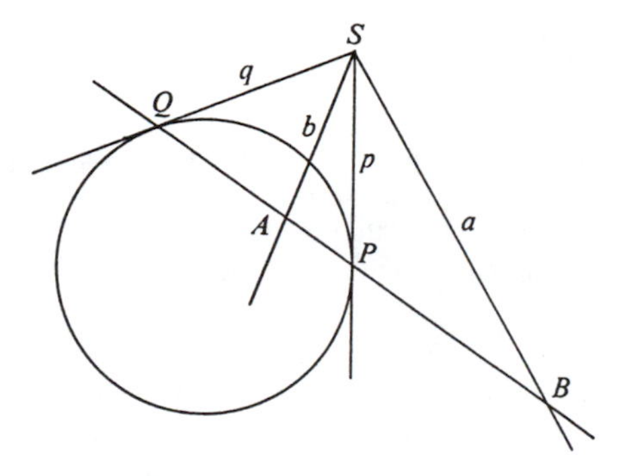

Figure 2.9d

2.9.10 THEOREM. *If A, B, C, D are four distinct collinear points, and a, b, c, d are their polars for a given circle, then $(AB,CD) = (ab,cd)$.*

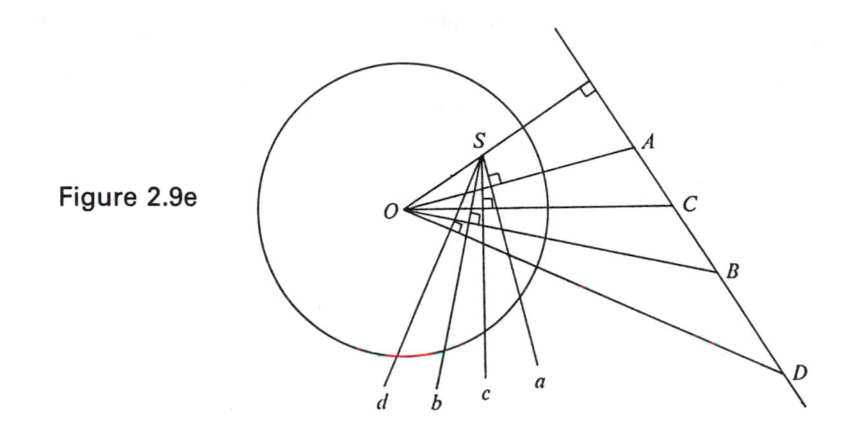

Figure 2.9e

Referring to Figure 2.9e, the polars a, b, c, d all pass through S, the pole of the line $ABCD$. Since each polar is perpendicular to the line joining its pole to the center O of the circle, we see that

$$(ab,cd) = O(AB,CD) = (AB,CD).$$

PROBLEMS

1. Establish Theorem 2.9.2.

2. Establish Theorem 2.9.6.

3. If P and Q are conjugate points for a circle, show that $(PQ)^2$ is equal to the sum of the powers of P and Q with respect to the circle.

4. If PR is a diameter of a circle K_1 orthogonal to a circle K_2 of center O, and if OP meets K_1 in Q, prove that line QR is the polar of P for K_2.

5. (a) If P and Q are conjugate points for circle K, prove that the circle on PQ as diameter is orthogonal to K.
   (b) If two circles are orthogonal, prove that the extremities of any diameter of one are conjugate points for the other.

6. (a) Let K_1, K_2, K_3 be three circles having a radical circle R, and let P be any point on R. Show that the polars of P for K_1, K_2, K_3 are concurrent.
   (b) A common tangent to two circles K_1 and K_2 touches them at P and Q respectively. Show that P and Q are conjugate points for any circle coaxial with K_1 and K_2.
   (c) The tangent to the circumcircle of triangle ABC at vertex A intersects line BC at T and is produced to U so that $AT = TU$. Prove that A and U are conjugate points for any circle passing through B and C.
   (d) Let ABC be a right triangle with right angle at B, and let B' be the midpoint of AC. Prove that A and C are conjugate points for any circle which touches BB' at B.
   (e) If a pair of opposite vertices of a square are conjugate points for a circle, prove that the other pair of opposite vertices are also conjugate points for the circle.

7. Prove *Salmon's Theorem*: If P, Q are two points, and PX, QY are the perpendiculars from P, Q to the polars of Q, P respectively, for a circle of center O, then $OP/OQ = PX/QY$.

2.10 APPLICATIONS OF RECIPROCATION

In this section we give a few applications of the reciprocation transformation. The first is a proof of Brianchon's Theorem for a circle (see Problem 4, Section 1.6). This theorem was discovered and published in a paper by C. J. Brianchon (1785–1864) in 1806, when still a student at the École Polytechnique in Paris, over 150 years after Pascal had stated his famous mystic hexagram theorem. Brianchon's paper was one of the first publications to employ the theory of poles and polars to obtain new geometrical results, and his theorem played a leading role in the recognition of the far-reaching principle of duality. The following proof of Brianchon's theorem is essentially that given by Brianchon himself.

2.10.1 BRIANCHON'S THEOREM FOR A CIRCLE. *If a hexagon (not necessarily convex) is circumscribed about a circle, the three lines joining pairs of opposite vertices are concurrent.*

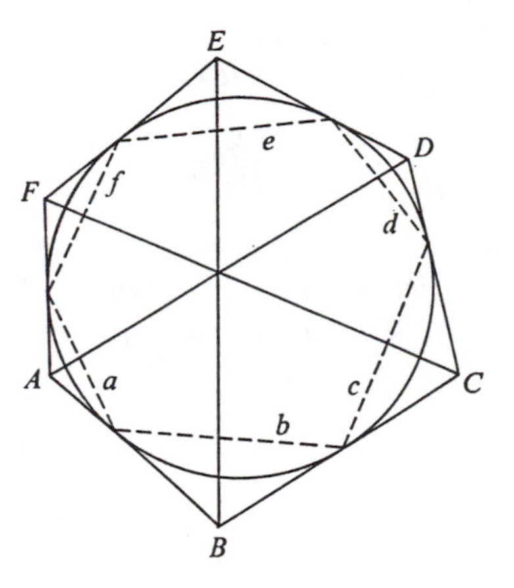

Figure 2.10a

Let $ABCDEF$ (see Figure 2.10a) be a hexagon circumscribed about a circle. The polars for the circle, of the vertices A, B, C, D, E, F, form the sides a, b, c, d, e, f of an inscribed hexagon. Since a is the polar of A and d the polar of D, the point ad is (by Theorem 2.9.3(3)) the pole of line AD. Similarly, the point be is the pole of line BE, and the point cf is the pole of line CF. Now, by Pascal's mystic hexagram theorem, the points ad, be, cf are collinear. It follows (by Theorem 2.9.4) that the polars AD, BE, CF are concurrent.

2.10.2 THEOREM. *Let ABCD be a complete quadrangle inscribed in a circle. Then each diagonal point of the quadrangle is the pole, for the circle, of the line determined by the other two diagonal points.*

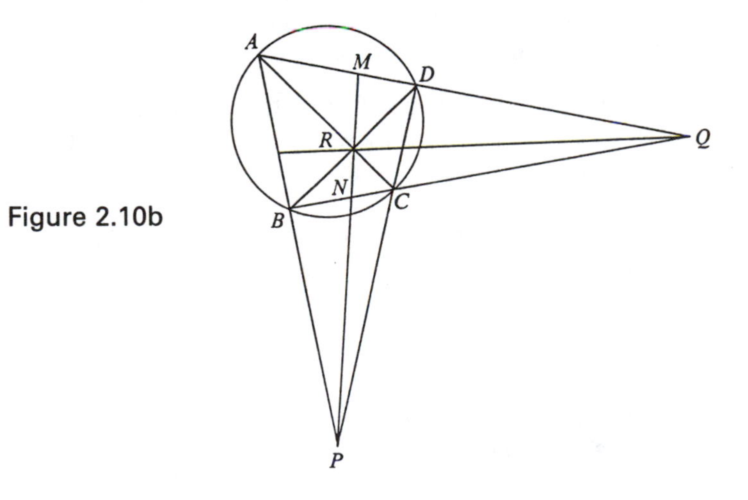

Figure 2.10b

For we have (see Figure 2.10b) $(AD,MQ) = (BC,NQ) = -1$, whence (by Corollary 2.9.8) line PR is the polar of point Q. Similarly line QR is the polar of point P. It then follows (by Theorem 2.9.3(3)) that line PQ is the polar of point R.

2.10.3 THE BUTTERFLY THEOREM. *Let O be the midpoint of a given chord of a circle, let two other chords TU and VW be drawn through O, and let TW and VU cut the given chord in E and F respectively. Then O is the midpoint of FE.*

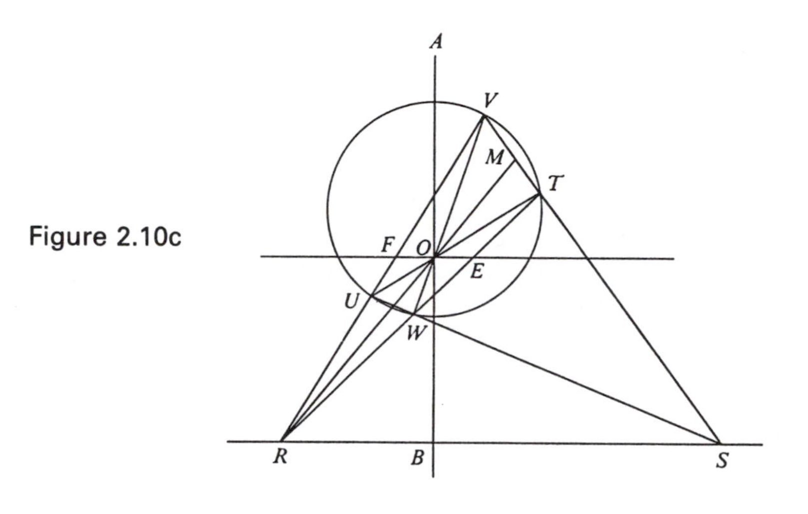

Figure 2.10c

If the given chord is a diameter of the circle, the theorem is obvious. Otherwise (see Figure 2.10c), produce TW and VU to intersect in R, and VT and UW to intersect in S. Then (by Theorem 2.10.2) RS is the polar of O, whence RS is perpendicular to the diametral line AOB. But FE is perpendicular to AOB. Therefore FE is parallel to RS. Now (by Theorem 1.8.11) $R(VT,MS) = -1$, whence $(FE,O\infty) = -1$, and O bisects FE.

Theorem 2.10.3 has received its name from the fancied resemblance of the figure of the theorem to a butterfly with outspread wings. It is a real stickler if one is limited to the use of only high school geometry.

2.10.4 THEOREM. *Let PQR be a triangle and let P', Q', R' be the poles of QR, RP, PQ for a circle K. Then P, Q, R are the poles of $Q'R'$, $R'P'$, $P'Q'$ for circle K.*

The easy proof is left to the reader.

2.10.5 DEFINITIONS. Two triangles are said to be *conjugate*, or *polar*, for a circle if each vertex of one triangle is the pole of a side of the other triangle. If the triangle is conjugate to itself—that is, each vertex is the pole of the opposite side—the triangle is said to be *self-conjugate*, or *self-polar*, for the circle.

We merely sketch the proofs of the next two theorems, leaving it to the reader to draw the accompanying figures and to complete details.

2.10.6 THEOREM. *Two triangles that are conjugate for a circle are copolar to one another.*

Let ABC and $A'B'C'$ be a pair of conjugate triangles. Let BC and $B'C'$ meet in M and let AA' cut BC in N and $B'C'$ in N'. Then (by Theorem 2.9.10) $(BC,NM) = A'(C'B',MN) = (B'C',N'M)$. It now follows that BB', CC', NN' are concurrent (by Corollary 1.6.2).

2.10.7 HESSE'S THEOREM FOR THE CIRCLE (1840). *If two pairs of opposite vertices of a complete quadrilateral are conjugate points for a circle, then the third pair of opposite vertices are also conjugate points for the circle.*

Let A, A'; B, B'; C, C' be the three pairs of opposite vertices of the quadrilateral, and suppose A, A'; B, B' are conjugate pairs of points for a circle. Let $A''B''C''$ be the triangle conjugate to triangle ABC. Now $B''C''$ passes through A', and $A''C''$ passes through B'. But triangles ABC, $A''B''C''$ are copolar (by Theorem 2.10.6), and therefore corresponding sides meet in three collinear points, two of which are seen to be A' and B'. It follows that C' must be the third point. That is, $B''A''$ passes through C', and C and C' are conjugate points.

PROBLEMS

1. A variable chord PQ of a given circle K passes through a fixed point T. Prove that the tangents at P and Q intersect on a fixed line t.

2. Assume the theorem: "The feet of the perpendiculars from any point on the circumcircle of a triangle to the sides of the triangle are collinear." (The line of collinearity is called the *Simson line* of the point for the triangle.)

   Let I be the incircle of a triangle ABC, and let perpendiculars through I to IA, IB, IC meet a given tangent to the incircle in P, Q, R. Prove that AP, BQ, CR are concurrent.

3. Consider the two propositions: (1) The lines joining the vertices of a triangle to the points of contact of the opposite sides with the incircle of the triangle are concurrent. (2) The tangents to the circumcircle of a triangle at the vertices of the triangle meet the opposite sides of the triangle in three collinear points.

   Show that Proposition (2) can be obtained from Proposition (1) by subjecting Proposition (1) to a reciprocation in the incircle of the triangle.

4. Draw a figure for Theorem 2.10.6 and verify on it the proof of the theorem given in the text.

5. If P, Q are two conjugate points for a circle, and R is the pole of PQ, show that PQR is a self-conjugate triangle.

6. If a triangle is self-conjugate for a circle, prove that its orthocenter is the center of the circle.

7. (a) Given an obtuse triangle, prove that there exists one and only one circle for which the triangle is self-conjugate. (The circle for which an obtuse triangle is self-conjugate is called the *polar circle* of the triangle.)
   (b) Prove that the polar circle of an obtuse triangle ABC has its center at the orthocenter of the triangle, and its radius equal to $[(\overline{HA})(\overline{HD})]^{1/2}$, where D is the foot of the altitude from A.
   (c) Prove that the inverse of the circumcircle of an obtuse triangle with respect to its polar circle is the nine-point circle of the triangle.

8. AB, CD are conjugate chords of a circle.
   (a) Show that $(AB,CD) = -1$.
   (b) Show that $(AC)(BD) = (BC)(AD) = (AB)(CD)/2$.

9. (a) Prove, in Figure 2.10b, that triangle PQR is self-conjugate for the circle $ABCD$.
   (b) Show that the circles on PQ, QR, RP as diameters are orthogonal to circle $ABCD$.

10. Draw a figure for Theorem 2.10.7 and verify on it the proof of the theorem given in the text.

11. Let X, Y, Z be the points of contact of the incircle of triangle ABC with the sides BC, CA, AB respectively, and let P be the point on the incircle diametrically opposite point X.
    (a) If A' is the midpoint of BC, show that AA', PX, YZ are concurrent.
    (b) If PA, PY, PZ cut line BC in M, Q, R respectively, show that $RM = MQ$.

2.11 SPACE TRANSFORMATIONS (OPTIONAL)

The elementary point transformations of unextended (three-dimensional) space are (1) translation, (2) rotation about an axis, (3) reflection in a point, (4) reflection in a line, (5) reflection in a plane, and (6) homothety. In (1), the set S of all points of unextended space is mapped onto itself by carrying each point P of S into a point P' of S such that $\overline{PP'}$ is equal and parallel to a given directed segment $\overline{AB}$ of space. There are no invariant points under a translation of nonzero vector $\overline{AB}$. In (2), each point P of S is carried into a point P' of S by rotating P about a fixed line in space through a given angle. The fixed line is called the axis of the rotation, and the points of the axis are the invariant points of the rotation.

In (3), each point P of S is carried into the point P' of S such that PP' is bisected by a fixed point O of space. The fixed point O is the only invariant point of the transformation. In (4), each point P of S is carried into the point P' of S such that PP' is perpendicularly bisected by a fixed line l of space. This transformation is obviously equivalent to a rotation of 180° about the line l. In (5), each point P of S is carried into the point P' of S such that PP' is perpendicularly bisected by a fixed plane p of space, and the points of p are the invariant points of the transformation. In (6), each point P of S is carried into the point P' of S collinear with P and a fixed point O of space, and such that $\overline{OP'}/\overline{OP} = k$, where k is a nonzero real number. If $k \neq 1$, the point O is the only invariant point of the transformation.

Certain compounds of the above elementary point transformations of space are basic in a study of similarities and isometries in space. These compounds are: (7) screw-displacement, (8) glide-reflection, (9) rotatory-reflection, and (10) space homology. A *screw-displacement* is the product of a rotation and a translation along the axis of rotation; a *glide-reflection* is the product of a reflection in a plane and a translation of vector $\overline{AB}$, where $\overline{AB}$ lies in the plane; a *rotatory-reflection* is the product of a reflection in a plane and a rotation about a fixed axis perpendicular to the plane; a *space homology* is the product of a homothety and a rotation about an axis passing through the center of the homothety.

Similarities and isometries in space are defined exactly as they were defined in a plane. Thus, a point transformation of unextended space onto itself that carries each pair of points A, B into a pair A', B' such that $A'B' = k(AB)$, where k is a fixed positive number, is called a *similarity*, and the particular case where $k = 1$ is called an *isometry*. A similarity is said to be *direct* or *opposite* according as tetrahedron $ABCD$ has or has not the same sense as tetrahedron $A'B'C'D'$.

We lack the space and time to develop the theory of similarities and isometries of space, and accordingly list the following interesting theorems without proof.

2.11.1 THEOREM. *Every isometry in space is the product of at most four reflections in planes.*

2.11.2 THEOREM. *Every isometry in space containing an invariant point is the product of at most three reflections in planes.*

2.11.3 THEOREM. *Every direct isometry in space is the product of two reflections in lines.*

2.11.4 THEOREM. *Any direct isometry is either a rotation, a translation, or a screw-displacement.*

2.11.5 THEOREM. *Any opposite isometry is either a rotatory-reflection or a glide-reflection.*

2.11.6 THEOREM. *Any nonisometric similarity is a space homology.*

The inversion transformation of Section 2.6 is also easily generalized to space.

2.11.7 DEFINITIONS AND NOTATION. We denote the sphere of center O and radius r by the symbol $O(r)$. If point P is not the center O of sphere $O(r)$, the *inverse* of P in, or with respect to, sphere $O(r)$ is the point P' lying on the line OP such that $(\overline{OP})(\overline{OP}') = r^2$. Sphere $O(r)$ is called the *sphere of inversion*, point O the *center of inversion*, r the *radius of inversion*, and r^2 the *power of inversion*.

2.11.8 CONVENTION AND DEFINITIONS. When working with inversion in space, we add to the set S of all points of space a single ideal *point at infinity*, to be considered as lying on every plane of space, and this ideal point, Z, shall be the image under the inversion of the center O of inversion, and the center O of inversion shall be the image under the inversion of the ideal point Z. Space, augmented in this way, will be referred to as *inversive space*.

It is now apparent that space inversion is a transformation of inversive space onto itself. Much of the theory of planar inversion can easily be extended to space inversion. For example, if "sphere" (with the quotation marks) denotes either a plane or a sphere, one can prove the following theorem.

2.11.9 THEOREM. *In a space inversion, "spheres" invert into "spheres," and "circles" invert into "circles."*

In particular, one can show that: (1) a plane through the center of inversion inverts into itself, (2) a plane not through the center of inversion inverts into a sphere through the center of inversion, (3) a sphere through the center of inversion inverts into a plane not through the center of inversion, (4) a sphere not through the center of inversion inverts into a sphere not through the center of inversion.

The pole-polar relation of Section 2.9 can be generalized to extended space.

2.11.10 DEFINITIONS. Let $O(r)$ be a fixed sphere and let P be any ordinary point other than O. Let P' be the inverse of P in the sphere $O(r)$. Then the plane p through P' and perpendicular to OPP' is called the *polar* of P for the sphere $O(r)$. The *polar of* O is taken as the plane at infinity, and the *polar of an ideal point* P is taken as the plane through O perpendicular to the direction OP. If plane p is the polar of point P, then point P is called the *pole* of plane p.

In the pole-polar relation of extended space we have a transformation of the set of all points of extended space onto the set of all planes of extended space. It can be shown that this transformation carries a range of points into a pencil of planes, and we accordingly have a straight line (considered as the base of a range of points) associated with a straight line (considered as the axis of a pencil of planes). This pole-polar transformation of extended space gives rise to a remarkable principle of duality of extended space.

Besides point transformations mapping a whole three-dimensional space onto itself, there are point transformations which map a part of space onto another part of space. Consider, for example, two planes p_1 and p_2, in unextended space, and a fixed direction not parallel to either plane. We may induce a transformation of the set of all points of p_1 onto the set of all points of p_2 by the simple procedure of associating with each point of P_1 of p_1 the point P_2 of p_2 such that P_1P_2 is parallel to the given direction. If we consider p_1 and p_2 as immersed in extended space, then the transformation carries the line at infinity of p_1 into the line at infinity of p_2. As another example, again consider two ordinary planes p_1 and p_2 in extended space, and a point O not on either plane. We can induce a transformation of the set of all points of p_1 onto the set of all points of p_2 by associating with each point P_1 of p_1 the point P_2 of p_2 such that O, P_1, P_2 are collinear.

We conclude the section with an outline of a point transformation of a part of inversive space onto another part of itself. The transformation, known as stereographic projection, was known to the ancient Greeks and affords a simple and useful method of transferring figures from a plane to the surface of a sphere, or vice versa.

2.11.11 DEFINITION AND NOTATION. Let K be a sphere of diameter d, and p a plane tangent to K at its south pole S; let N be the north pole of K. If P is any point of the sphere other than N, we associate with it the point P' of p such that N, P, P' are collinear. With N we associate the point at infinity on the inversive plane p. This transformation of the set of all points of the sphere K onto the set of all points of the inversive plane p is called *stereographic projection*.

The proofs of the first two of the following theorems are left to the reader.

2.11.12 THEOREM. *The meridians on K correspond to the straight lines* *on p through S, and the circles of latitude on K correspond to the circles on p having center S. In particular, the equator of K corresponds to the circle on p of center S and radius d.*

2.11.13 THEOREM. *The circles on K through N correspond to the straight lines on p.*

2.11.14 THEOREM. *If points P and Q on the sphere K are reflections of one another in the equatorial plane of K, then their images P′, Q′ on plane p are inverses of one another in the circle S(d) of p.*

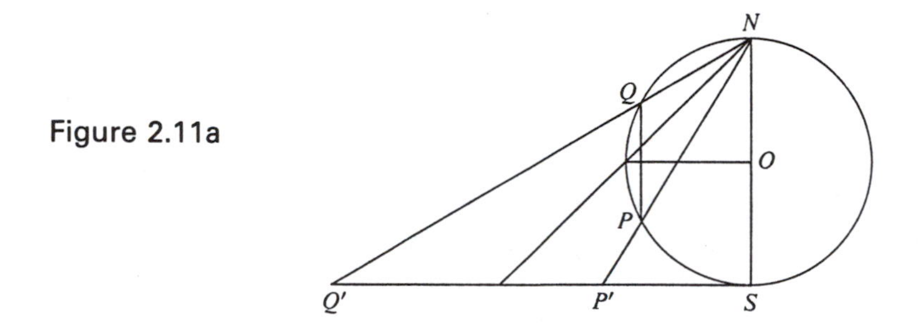

Figure 2.11a

For (see Figure 2.11a) $SP' = d \tan PNS$, $SQ' = d \tan QNS$. But angles PNS and QNS are complementary, whence $(SP')(SQ') = d^2$. Clearly S, P', Q' are collinear.

Theorem 2.11.14 exhibits an interesting interpretation of a planar inversion. To perform a planar inversion, we may first project the figure stereographically onto a sphere, then interchange the hemispheres by reflection in the equatorial plane, and finally project stereographically back onto the plane.

2.11.15 THEOREM. *Stereographic projection is a space inversion in the sphere N(d).*

For (see Figure 2.11a) $NP = d \cos PNS$, $NP' = d \sec PNS$, whence $(NP)(NP') = d^2$, and of course N, P, P' are collinear.

2.11.16 THEOREM. *Circles on K correspond to "circles" on p.*

This follows from Theorems 2.11.15 and 2.11.9.

2.11.17 THEOREM. *The angle between two lines of p is equal to the angle between their stereographic projections on K.*

Two lines AB, AC of p map into circles $NA'B'$, $NA'C'$ of K. The tangents to these circles at N are parallel to AB and AC respectively. But the circles intersect at equal angles at N and A'. It follows that an angle between the circles at A' is equal to the corresponding angle between the straight lines at A.

Considered as a mapping of the sphere onto the plane, stereographic projection furnishes a satisfactory representation on a plane of a limited region of the sphere. In fact, stereographic projection is one of the most commonly used methods of constructing geographical maps. Particularly useful in such maps is the preservation of angles guaranteed by Theorem 2.11.17.

Stereographic projection also furnishes an elegant and timesaving way of obtaining the formulas of spherical trigonometry from those of plane trigonometry. This approach to spherical trigonometry was developed by the crystallographer and mineralogist Giuseppe Cesàro (1849–1939). For an exposition in English see J. D. H. Donnay, *Spherical Trigonometry after the Cesàro Method* (New York: Interscience Publishers, Inc., 1945).

In the Appendix will be found a brief treatment of another useful and attractive transformation. Though, strictly speaking, this transformation is not a transformation of modern elementary geometry, but rather a transformation of projective geometry, many instructors like to include it, for it furnishes still another way of establishing many theorems of modern elementary geometry.

PROBLEMS

1. (a) Which of the elementary point transformations of unextended space are involutoric? (b) Which are direct isometries? (c) Which are opposite isometries?

2. (a) Are the rotation and the translation of a screw-displacement commutative?
   (b) Are the reflection and the translation of a glide-reflection commutative?
   (c) Are the reflection and the rotation of a rotatory-reflection commutative?

3. (a) Give a proof of Theorem 2.11.1 patterned after that of Theorem 2.4.3.
   (b) Give a proof of Theorem 2.11.2 patterned after that of Theorem 2.4.4.

4. State a space analogue of Theorem 2.4.2.

5. (a) Prove that a translation in space can be factored into a product of reflections in two parallel planes.
   (b) Prove that a rotation about an axis can be factored into a product of reflections in two planes passing through the axis of rotation.

6. Show that the product of reflections in two perpendicular planes is a reflection in the line of intersection of the two planes.

7. Show that if a direct isometry leaves a point O fixed, it leaves some line through O fixed, that is, the isometry is a rotation.

8. Prove *Euler's Theorem on Rotations* (1776): The product of two rotations about axes through a point O is a rotation about an axis through O.

9. (a) Show that the product of two rotations of 180° about two given intersecting lines that form an angle θ is a rotation of 2θ about a line perpendicular to the two given lines at their point of intersection.
   (b) Show that the product of rotations of 180° about three mutually perpendicular concurrent lines is the identity.

10. In a space homology H let k denote the ratio of homothety and θ the angle of rotation. Show that:
    (a) If $\theta = 0$, $k = 1$, then H is the identity.
    (b) If $\theta = 180°$, $k = 1$, then H is a reflection in a line.
    (c) If $\theta = \theta$, $k = 1$, then H is a rotation about an axis.
    (d) If $\theta = 0$, $k = -1$, then H is a reflection in a point.
    (e) If $\theta = 180°$, $k = -1$, then H is a reflection in a plane.
    (f) If $\theta = \theta$, $k = -1$, then H is a rotatory-reflection.
    (g) If $\theta = 0$, $k = k$, then H is a homothety.

11. Show that a "sphere" orthogonal to the sphere of inversion inverts into itself.

12. Show that a sphere through a pair of inverse points is orthogonal to the sphere of inversion.

13. Prove Theorem 2.11.9.

14. Given a tetrahedron $ABCD$ and a point M, prove that the tangent planes, at M, to the four spheres $MBCD$, $MCDA$, $MDAB$, $MABC$, meet the respective faces BCD, CDA, DAB, ABC in four coplanar lines. (This is Problem E 493 of *The American Mathematical Monthly*, June–July 1942.)

15. Prove *Frederick Soddy's Hexlet Theorem* (1936): Let S_1, S_2, S_3 be three spheres all touching one another. Let $K_1, K_2, \ldots$ be a sequence of spheres touching one another successively and all touching S_1, S_2, S_3. Show that K_6 touches K_1.

16. (a) Show that if, for a given sphere, the polar plane of point P passes through point Q, then the polar plane of point Q passes through point P.
    (b) Show that if, for a given sphere, the pole of plane p lies on plane q, the pole of q lies on p.
    (c) Show that if, for a given sphere, the noncollinear points P, Q, R are the poles of p, q, r, then the pole of plane PQR is the point of intersection of p, q, r.

17. Show that the polars, for a given sphere, of a range of points form a pencil of planes.

18. Define *conjugate points* and *conjugate planes* for a sphere.

19. Define *conjugate tetrahedra* and *self-conjugate tetrahedron* of a sphere.

20. (a) Establish Theorem 2.11.12.
    (b) Establish Theorem 2.11.13.

BIBLIOGRAPHY

See the Bibliography of Chapter 1; also:

BARRY, E. H., *Introduction to Geometrical Transformations*. Boston, Mass.: Prindle, Weber & Schmidt, 1966.

CHOQUET, GUSTAVE, *Geometry in a Modern Setting*. Boston, Mass.: Houghton Mifflin Company, 1969.

COXETER, H. S. M., *Introduction to Geometry*. New York: John Wiley and Sons, 1961.

COXFORD, A. F., and Z. P. USISKIN, *Geometry, a Transformation Approach*. River Forest, Ill.: Laidlaw Brothers, 1971.

DONNAY, J. D. H., *Spherical Trigonometry after the Cesàro Method*. New York: Interscience Publishers, Inc., 1945.

ECCLES, F. M., *An Introduction to Transformational Geometry*. Reading, Mass.: Addison-Wesley, 1971.

FORDER, H. G., *Geometry*. New York: Hutchinson's University Library, 1950.

GANS, DAVID, *Transformations and Geometries*. New York: Appleton-Century-Crofts, 1969.

GRAUSTEIN, W. C., *Introduction to Higher Geometry*. New York: The Macmillan Company, 1930.

JEGER, MAX, *Transformation Geometry*, tr. by A. W. Deicke and A. G. Howson. London: George Allen and Unwin, 1964.

KLEIN, FELIX, *Elementary Mathematics from an Advanced Standpoint, Geometry*, tr. by E. R. Hedrick and C. A. Noble. New York: Dover Publications, 1939. First German edition published in 1908.

LEVI, HOWARD, *Topics in Geometry*. Boston, Mass.: Prindle, Weber & Schmidt, 1968.

LEVY, L. S., *Geometry: Modern Mathematics via the Euclidean Plane*. Boston, Mass.: Prindle, Weber & Schmidt, 1970.

MODENOV, P. S., and A. S. PARKHOMENKO, *Geometric Transformations* (2 vols.), tr. by M. B. P. Slater. New York: Academic Press, 1965.

SMART, J. R., *Modern Geometries*, 3d ed. Pacific Grove, Calif.: Brooks/Cole Publishing Company, 1988.

YAGLOM, I. M., *Geometric Transformations*, tr. by Allen Shields. New York: Random House, New Mathematical Library, No. 8, 1962.

3

Euclidean Constructions

There is much to be said in favor of a game that you play alone. It can be played or abandoned whenever you wish. There is no bother about securing a willing and suitable opponent, nor do you annoy anyone if you suddenly decide to desist play. Since you are, in a sense, your own opponent, the company is most congenial and perfectly matched in skill and intelligence, and there is no embarrassing sarcastic utterance should you make a stupid play. The game is particularly good if it is truly challenging and if it possesses manifold variety. It is still better if the rules of the game are very few and simple. And little more can be asked if, in addition, the game requires no highly specialized equipment, and so can be played almost anywhere and at almost any time.

The Greek geometers of antiquity devised a game—we might call it *geometrical solitaire*—which, judged on all the above points, must surely stand at the very top of any list of games to be played alone. Over the ages it has attracted hosts of players, and though now well over 2000 years old, it seems not to have lost any of its singular charm and appeal. This chapter is concerned with a few facets of this fascinating game, and with some of its interesting modern variants.

3.1 THE EUCLIDEAN TOOLS

For convenience we repeat here the first three postulates of Euclid's *Elements*:

1. A straight line can be drawn from any point to any point.
2. A finite straight line can be produced continuously in a straight line.
3. A circle may be described with any center and distance.

These postulates state the primitive constructions from which all other constructions in the *Elements* are to be compounded. They constitute, so to speak, the rules of the game of Euclidean construction. Since they restrict constructions to only those that can be made in a permissible way with straightedge and compass,* these two instruments, so limited, are known as the *Euclidean tools*.

The first two postulates tell us what we can do with a Euclidean straightedge; we are permitted to draw as much as may be desired of the straight line determined by any two given points. The third postulate tells us what we can do with the Euclidean compass; we are permitted to draw the circle of given center and having any straight line segment radiating from that center as a radius—that is, we are permitted to draw the circle of given center and passing through a given point. Note that neither instrument is to be used for transferring distances. This means that the straightedge cannot be marked, and the compass must be regarded as having the characteristic that if either leg is lifted from the paper, the instrument immediately collapses. For this reason, a Euclidean compass is often referred to as a *collapsing compass*; it differs from a *modern compass*, which retains its opening and hence can be used as a divider for transferring distances.

It would seem that a modern compass might be more powerful than a collapsing compass. Curiously enough, such turns out not to be the case; any construction performable with a modern compass can also be carried out (in perhaps a longer way) by means of a collapsing compass. We prove this fact as our first theorem, right after introducing the following convenient notation.

3.1.1 NOTATION. The circle with center O and passing through a given point C will be denoted by $O(C)$, and the circle with center O and radius equal to a given segment AB will be denoted by $O(AB)$.

3.1.2 THEOREM. *The collapsing and modern compasses are equivalent.*

To prove the theorem it suffices to show that we may, with a collapsing compass, construct any circle $O(AB)$. This may be accomplished as follows

* Though contrary to common English usage, we shall use this word in the singular.

Figure 3.1a

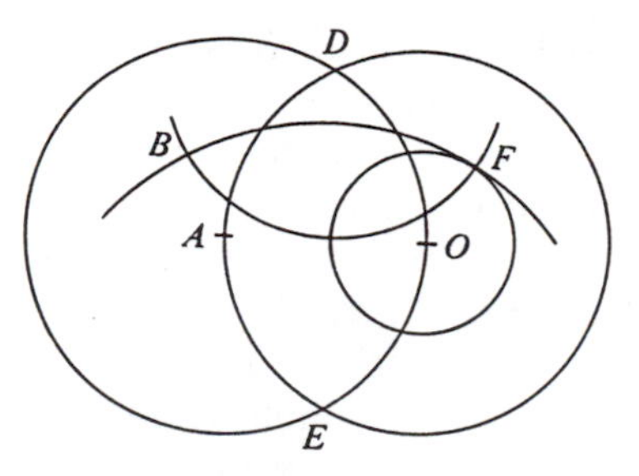

(see Figure 3.1a). Draw circles $A(O)$ and $O(A)$ to intersect in D and E; draw circles $D(B)$ and $E(B)$ to intersect again in F; draw circle $O(F)$. It is an easy matter to prove that $OF = AB$, whence circle $O(F)$ is the same as circle $O(AB)$.

In view of Theorem 3.1.2, we may dispense with the Euclidean, or collapsing compass, and in its place employ the more convenient modern compass. We are assured that the set of constructions performable with straightedge and Euclidean compass is the same as the set performable with straightedge and modern compass. As a matter of fact, in all our construction work, we shall not be interested in actually and exactly carrying out the constructions, but merely in assuring ourselves that such constructions are possible. To use a phrase of Jacob Steiner, we shall do our constructions "simply by means of the tongue," rather than with actual instruments on paper. We seek then, at least for the time being, the construction easiest to describe rather than the simplest or best construction to actually carry out with the instruments.

If one were asked to find the midpoint of a given line segment using only the straightedge, one would be justified in exclaiming that surely the Euclidean straightedge alone will not suffice, and that some additional tool or permission must be furnished. The same is true of the combined Euclidean tools; there are constructions that cannot be performed with these tools alone, at least under the restrictions imposed upon them. Three famous problems of this sort, which originated in ancient Greece, are

1. *The duplication of the cube*, or the problem of constructing the edge of a cube having twice the volume of a given cube.
2. *The trisection of an angle*, or the problem of dividing a given arbitrary angle into three equal parts.
3. *The quadrature of the circle*, or the problem of constructing a square having an area equal to that of a given circle.

The fact that there are constructions beyond the Euclidean tools adds a certain zest to the construction game. It becomes desirable to obtain a criterion for determining whether a given construction problem is or is not within the power of our tools. Synthetic geometry has not been able to cope with this problem, and accordingly a satisfactory discussion of this interesting facet of Euclidean constructions cannot be given here but must be deferred to some other place where the needed algebraic background can be developed. Nevertheless, we shall briefly return to the matter in our concluding Section 3.9.

But, in spite of the limited power of our instruments, really intricate constructions can be accomplished with them. Thus, though with our instruments we cannot, for example, solve the seemingly simple problem of drawing the two lines trisecting an angle of 60°, we can draw all the circles that touch three given circles (the *problem of Apollonius*); we can draw three circles in the angles of a triangle such that each circle touches the other two and also the two sides of the angle (the *problem of Malfatti*); we can inscribe in a given circle a triangle whose sides, produced if necessary, pass through three given points (the *Castillon-Cramer problem*).

As a concluding remark of this section we point out that Euclid used constructions in the sense of existence theorems—to prove that certain entities actually exist. Thus one may define a *bisector* of a given angle as a line in the plane of the angle, passing through the vertex of the angle, and such that it divides the given angle into two equal angles. But a definition does not establish the existence of the thing being defined; this requires proof. To show that a given angle does possess a bisector, we show that this entity can actually be constructed. Existence theorems are very important in mathematics, and actual construction of an entity is the most satisfying way of proving its existence. One might define a *square circle* as a figure that is both a square and a circle, but one would never be able to prove that such an entity exists; the class of square circles is a class without any members. In mathematics it is nice to know that the set of entities satisfying a certain definition is not just the empty set.

PROBLEMS

1. In the proof of Theorem 3.1.2, show that $OF = AB$.

2. A student reading Euclid's *Elements* for the first time might experience surprise at the three opening propositions of Book I (for statements of these propositions see Appendix 1). These three propositions are constructions, and are trivial with straightedge and *modern* compass, but require some ingenuity with straightedge and *Euclidean* compass.
   (a) Solve I 1 with Euclidean tools.
   (b) Solve I 2 with Euclidean tools.
   (c) Solve I 3 with Euclidean tools.
   (d) Show that I 2 proves that the straightedge and Euclidean compass are equivalent to the straightedge and modern compass.

3. Consider the following two arguments:

I. Proposition. *Of all triangles inscribed in a circle, the equilateral is the greatest.*

1. If ABC is a nonequilateral triangle inscribed in a circle, so that $AB \neq AC$, say, construct triangle XBC, where X is the intersection of the perpendicular bisector of BC with arc BAC.
2. Then triangle $XBC >$ triangle ABC.
3. Hence, if we have a nonequilateral triangle inscribed in a circle, we can always construct a greater inscribed triangle.
4. Therefore, of all triangles inscribed in a circle, the equilateral is the greatest.

II. Proposition. *Of all natural numbers, 1 is the greatest.*

1. If m is a natural number other than 1, construct the natural number m^2.
2. Then $m^2 > m$.
3. Hence, if we have a natural number other than 1, we can always construct a greater natural number.
4. Therefore, of all natural numbers, 1 is the greatest.

Now the conclusion in argument I is true, and that in argument II is false. But the two arguments are formally identical. What, then, is wrong?

3.2. THE METHOD OF LOCI

In this section we very briefly consider what is perhaps the most basic method in the solution of geometric construction problems. It can be used alone or in combination with some other method. It may be considered as a fundamental maneuver in the construction game.

The solution of a construction problem very often depends upon first finding some key point. Thus the problem of drawing a circle through three given points is essentially solved once the center of the circle is located. Again, the problem of drawing a tangent to a circle from an external point is essentially solved once the point of contact of the tangent with the circle has been found. Now the key point satisfies certain conditions, and each condition considered alone generally restricts the position of the key point to a certain locus. The key point is thus found at the intersections of certain loci. This method of solving a construction problem is aptly referred to as the *method of loci.*

To illustrate, denote the three given points in our first problem above by A, B, C. Now the sought center O of the circle through A, B, C must be equidistant from A and B and also from B and C. The first condition places O on the perpendicular bisector of AB, and the second condition places O on the perpendicular bisector of BC. The point O is thus found at the intersection, if it exists, of these two perpendicular bisectors. If the three given points are not collinear, there is exactly one solution; otherwise there is none.

Suppose, in our second problem above, we denote the center of the given circle by O, the external point by E, and the sought point of contact of the tangent from E to the circle by T. Now T, first of all, lies on the given circle. Also, since $\sphericalangle OTE = 90°$, T lies on the circle having OE as diameter. The sought point T is thus found at an intersection of these two circles. There are always two solutions to the problem.

In order to apply the method of loci to the solution of geometric constructions, it is evidently of great value to know a considerable number of loci that are constructible straight lines and circles. Here are a few such loci.

1. The locus of points at a given distance from a given point is the circle having the given point as center and the given distance as radius.
2. The locus of points at a given distance from a given line consists of the two lines parallel to the given line and at the given distance from it.
3. The locus of points equidistant from two given points is the perpendicular bisector of the segment joining the two given points.
4. The locus of points equidistant from two given intersecting lines consists of the bisectors of the angles formed by the two given lines.
5. The locus of points from which lines drawn to the endpoints of a given line segment enclose a given angle consists of a pair of congruent circular arcs having the given segment as a chord, the two arcs lying on opposite sides of the given segment. In particular, if the given angle is a right angle, the two arcs are the semicircles having the given segment as diameter.
6. The locus of points whose distances from two given points A and B have a given ratio $k \neq 1$ is the circle on IE as diameter, where I and E divide AB internally and externally in the given ratio. (This is the circle of Apollonius for the given segment and the given ratio.)
7. The locus of points whose distances from two given intersecting lines have a given ratio k is a pair of straight lines through the point of intersection of the given lines. Locus (4) is the special case where $k = 1$.
8. The locus of points for which the difference of the squares of the distances from two given points is a constant is a straight line perpendicular to the line determined by the two given points.
9. The locus of points for which the sum of the squares of the distances from two given points is a constant is a circle having its center at the midpoint of the segment joining the two given points.

The reader should not only verify the correctness of each of the above loci, but also make sure that each one can actually be constructed with compass and straightedge. For example, to construct locus (5) the reader might lay off the supplement of the given angle at one end A of the given segment AB and then find the center of one of the desired arcs as the intersection of the perpendicular bisector of AB and the perpendicular at A to the other side of the layed-off angle. Locus (6) is easily constructed once the points I and E are found, and the finding of these was the subject matter of Problem 7, Section 1.1. Each of the lines of locus (7) can be found once one point, other than the intersection of the two given lines, has been found, and such a point can be found by (2), any two distances having the given ratio being chosen. To construct locus (8), let the given points be A and B and denote the given difference of squares of distances by d^2. Construct any right triangle having a leg d and the other leg greater than half of AB. Using A and B as centers and radii equal to the hypotenuse and other leg, respectively, of the constructed right triangle, find a point P on the sought locus. We are here assuming that P has its greater distance from A. To find the points M' and N' where locus (9) cuts the segment joining the two given points A and B, draw angle $BAD = 45°$ and cut line AD in M and N with circle $B(s)$, where s^2 is the given sum of squares of distances. Then M' and N' are the feet of the perpendiculars from M and N on AB.

Let us now consider a few construction problems solvable by the method of loci.

3.2.1 PROBLEM. *Draw a circle passing through two given points and subtending a given angle at a third point.*

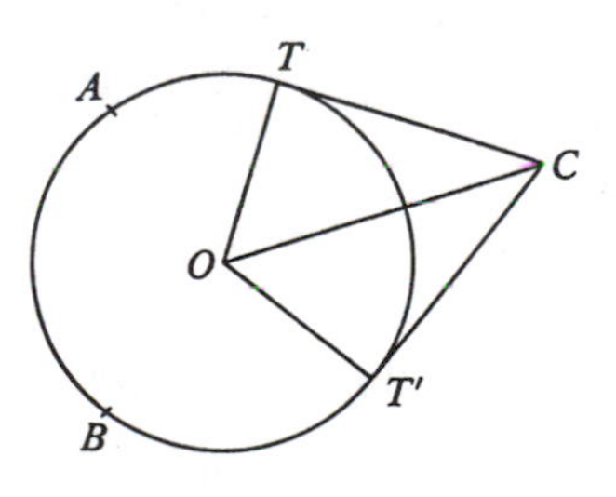

Figure 3.2a

Referring to Figure 3.2a, let A and B be the two given points and let C be the third point. Denote the center of the sought circle by O and let T and T' denote the points of contact of the tangents to the circle from point C. Since $\sphericalangle TCT'$ is given, the form of right triangle OTC is known. That is, we know the ratio OT/OC. But $OA/OC = OB/OC = OT/OC$. It follows that O lies on the circle of Apollonius for A and C and ratio OT/OC, and on the circle of Apollonius for B and C and ratio OT/OC. Point O is thus found at the intersections, if any exist, of these two circles of Apollonius. The details are left to the reader.

3.2.2 Problem. *Find a point whose distances from three given lines have given ratios.*

Use locus (7).

3.2.3 Problem. *In a triangle find a point whose distances from the three vertices have given ratios.*

Use locus (6).

3.2.4 Problem. *Construct a triangle given one side, the altitude on that side, and the sum of the squares of the other two sides.*

Find the vertex opposite the given side by using loci (2) and (9).

3.2.5 Problem. *Construct a triangle given one side, the opposite angle, and the difference of the squares of the other two sides.*

Find the vertex opposite the given side by using loci (5) and (8).

PROBLEMS

1. Establish the constructions given in the text for loci (1) through (9).
2. Complete the details of Problem 3.2.1.
3. Construct a triangle given one side and the altitude and median to that side.
4. Draw a circle touching two given parallel lines and passing through a given point.
5. Construct a triangle given one side, the opposite angle, and the median to the given side.
6. Find a point at which three given circles subtend equal angles.
7. Two balls are placed on a diameter of a circular billiard table. How must one ball be played in order to hit the other after its recoil from the circumference?
8. Through two given points of a circle draw two parallel chords whose sum shall have a given length.
9. Draw a circle of given radius touching a given circle and having its center on a given line.
10. Inscribe a right triangle in a given circle so that each leg will pass through a given point.
11. Draw a circle of given radius, passing through a given point, and cutting off a chord of given length on a given line.

12. Draw a circle tangent to a given line at a given point and also tangent to a given circle.

13. Draw a tangent to a given circle so that a given line cuts off on the tangent a given distance from the point of contact.

14. Construct a cyclic quadrilateral given one angle, an adjacent side, and the two diagonals.

15. Through a given point draw a line intersecting a given circle so that the distances of the points of intersection from a given line have a given sum.

16. Find a point from which three parts AB, BC, CD of a given line are seen under equal angles.

17. Find the locus of points for which the distances from two given lines have a given sum.

18. (a) On the circumference of a given circle find a point for which the sum of the distances from two given lines is given.
(b) Find the point on the given circle for which the sum of the distances from the two given lines is a minimum.

3.3 THE METHOD OF TRANSFORMATION

There are many geometric construction problems that can be solved by applying one of the transformations discussed in the previous chapter. We illustrate this *method of transformation* by the following sequence of problems, in which certain details are left to the reader.

3.3.1 PROBLEM. *Draw a line in a given direction on which two given circles cut off equal chords.*

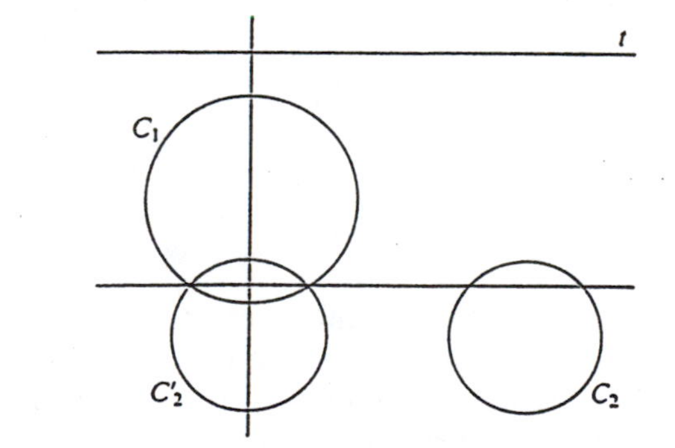

Figure 3.3a

In Figure 3.3a, let C_1 and C_2 be the given circles and let t be a line in the given direction. Translate C_2 parallel to t to position C_2' in which the line of centers of C_1 and C_2' is perpendicular to t. Then the line through the points of intersection, if such exist, of C_1 and C_2' is the sought line.

3.3.2 Problem. *Through one of the points of intersection of two given intersecting circles, draw a line on which the two circles cut off equal chords.*

Figure 3.3b

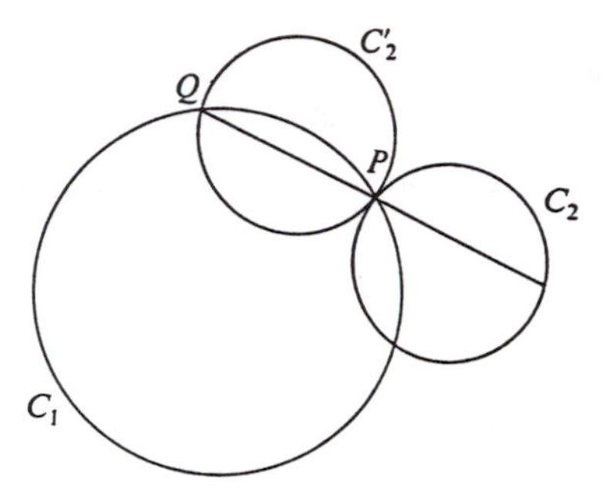

In Figure 3.3b, let C_1 and C_2 be the two given intersecting circles, and let P be one of the points of intersection. Reflect C_2 in point P into position C_2' and let Q be the other intersection of C_1 and C_2'. Then QP is the sought line.

3.3.3 Problem. *Inscribe a square in a given triangle, so that one side of the square lies on a given side of the triangle.*

Figure 3.3c

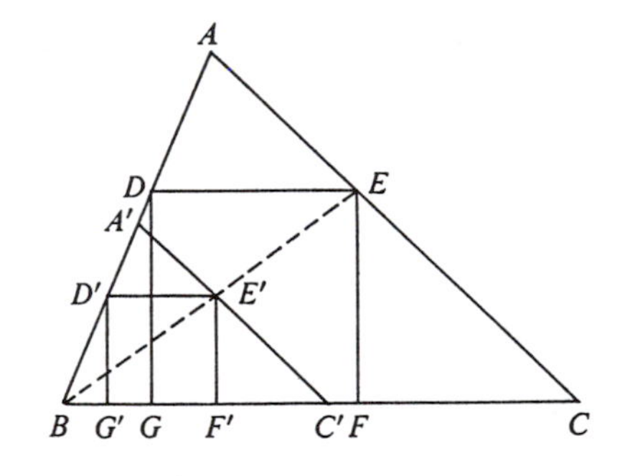

In Figure 3.3c, let ABC be the given triangle and BC the side on which the required square is to lie. Choose any point D' on side AB and construct the square $D'E'F'G'$ as indicated in the figure. If E' falls on AC, the problem is solved. Otherwise we have solved the problem for a triangle $A'BC'$ which is homothetic to triangle ABC with B as center of homothety. It follows that line BE' cuts AC in vertex E of the sought square inscribed in triangle ABC.

It is interesting, in connection with the last problem, to contemplate that from a sheer guess of the position of the sought square we were able to find the actual position of the square. The method employed, often called the *method of similitude*, is a geometric counterpart of the *rule of false position* used by the ancient Egyptians to solve linear equations in one unknown. Suppose, for example, we are to solve the simple equation $x + x/5 = 24$. Assume any convenient value of x, say $x = 5$. Then $x + x/5 = 6$, instead of 24. Since 6 must be multiplied by 4 to give the required 24, the correct value of x must be 4(5), or 20. From a sheer guess, and without employment of algebraic procedures, we have obtained the correct answer.

3.3.4 A GENERAL PROBLEM. *Given a point O and two curves C_1 and C_2, locate a triangle OP_1P_2, where P_1 is on C_1 and P_2 is on C_2, similar to a given triangle $O'P_1'P_2'$.*

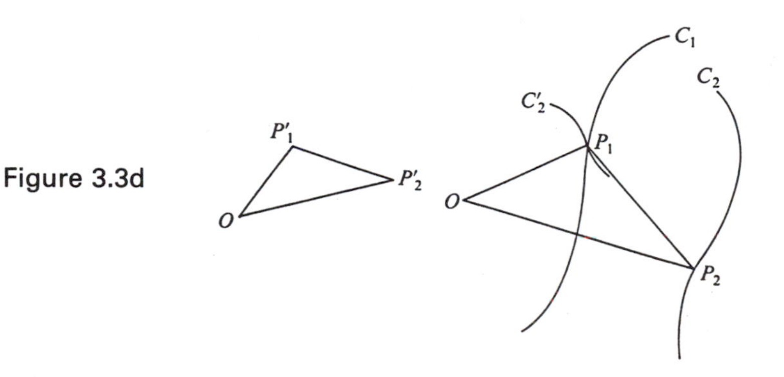

Figure 3.3d

In Figure 3.3d, let C_2' be the map of C_2 under the homology

$$H(O, \sphericalangle P_2'O'P_1', O'P_1'/O'P_2').$$

Then C_1 and C_2' intersect in the possible positions of P_1. If C_1 and C_2 are "circles," the problem can be solved with Euclidean tools.

3.3.5 PROBLEM. *Draw an equilateral triangle having its three vertices on three given parallel lines.*

In Figure 3.3e, choose any point O on one of the three given parallel lines, and denote the other two parallel lines by C_1 and C_2. We may now apply the above General Problem 3.3.4 by subjecting line C_2 to the homology $H(O, 60°, 1)$.

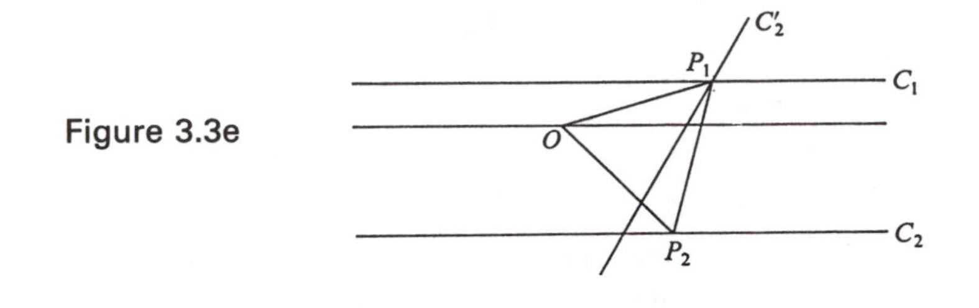

Figure 3.3e

3.3.6 PROBLEM. *Draw a "circle" touching three given concurrent non-coaxial circles.*

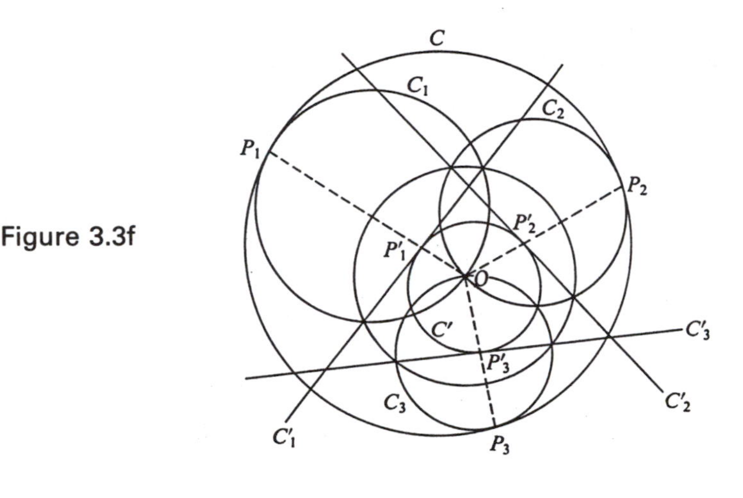

Figure 3.3f

In Figure 3.3f, let the three given circles C_1, C_2, C_3 intersect in point O. Subject the figure to any convenient inversion of center O. Then C_1, C_2, C_3 become three straight lines C_1', C_2', C_3', which are easily constructed with Euclidean tools (they are actually the common chords of C_1, C_2, C_3 with the circle of inversion). Draw a circle C' touching all three of the lines C_1', C_2', C_3'. The inverse C of this circle, which can be constructed with Euclidean tools (if C' touches C_1', C_2', C_3' at P_1', P_2', P_3', respectively, then C touches C_1, C_2, C_3 at the points P_1, P_2, P_3 where OP_1', OP_2', OP_3' cut C_1, C_2, C_3 again), is a "circle" touching C_1, C_2, C_3. Note that there are four solutions to the problem.

The *method of homology*, illustrated in Problem 3.3.5, and the *method of inversion*, illustrated in Problem 3.3.6, are powerful methods, and many construction problems that would otherwise be very difficult yield to these methods. They are, of course, instances of the general method of transformation. Note that Problem 3.3.6 is a special case of the Problem of Apollonius.

PROBLEMS

1. Place a line segment equal and parallel to a given line segment and having its extremities on two given circles.

2. From a vessel two known points are seen under a given angle. The vessel sails a given distance in a known direction, and now the same two points are seen under another known angle. Find the position of the vessel.

3. In a given quadrilateral inscribe a parallelogram whose center is at a given point.

4. Solve the general problem: To a given line draw a perpendicular on which two given curves will cut off equal lengths measured from the foot of the perpendicular.

5. Place a square with two opposite vertices on a given line and the other two vertices on two given circles.

6. Draw a triangle given the positions of three points that divide the three sides in given ratios.

7. Inscribe a quadrilateral of given shape in a semicircle, a specified side of the quadrilateral lying along the diameter of the semicircle.

8. Given the focus and directrix of a parabola, find the points of intersection of the parabola with a given line.

9. Draw a circle passing through a given point and touching two given lines.

10. Find points D and E on sides AB and AC of a triangle ABC so that $BD = DE = EC$.

11. Draw a triangle given A, $a + b$, $a + c$.

12. Solve the general problem: Through a given point O draw a line intersecting two given curves C_1 and C_2 in points P_1 and P_2 so that OP_1 and OP_2 shall be in a given ratio to one another.

13. Through a given point O within a circle draw a chord that is divided by O in a given ratio.

14. Through a given point O on a given circle draw a chord that is bisected by another given chord.

15. In a given triangle ABC inscribe another, $A'B'C'$, which shall have its sides parallel to three given lines.

16. Two radii are drawn in a circle. Draw a chord that will be trisected by the radii.

17. In a given parallelogram inscribe an isosceles triangle of given vertex angle and with its vertex at a given vertex of the parallelogram.

18. Draw an equilateral triangle having its three vertices on three given concentric circles.

19. Solve the general problem: Through a given point O draw a line cutting two given curves C_1 and C_2 in points P_1 and P_2 such that $(OP_1)(OP_2)$ is a given constant.

20. Through a given point O draw a line cutting two given lines in points A and B so that $(OA)(OB)$ is given.

21. Through a point of intersection of two circles draw a line on which the two circles intercept chords having a given product.

22. Draw a circle passing through a given point P and touching two given circles.

23. Draw a circle through two given points and tangent to a given circle.

24. Draw a circle tangent externally to three given mutually external circles.

3.4. THE DOUBLE POINTS OF TWO COAXIAL HOMOGRAPHIC RANGES (OPTIONAL)

In this section we describe another, and very clever, maneuver in the construction game. Some writers call this maneuver the *method of trial and error*. We need the following basic construction.

3.4.1 PROBLEM. *Find the common, or double, points of two distinct coaxial homographic ranges.*

A number of ingenious solutions have been devised for this problem. We give one here which is based upon simple cross-ratio properties of a circle. In Figure 3.4a, let A and A', B and B', C and C' be three pairs of corresponding points of two distinct coaxial homographic ranges. Draw any circle in the plane and on it take any point O. Draw the six lines OA, OB, OC, OA', OB', OC' and let them cut the circle again in A_1, B_1, C_1,

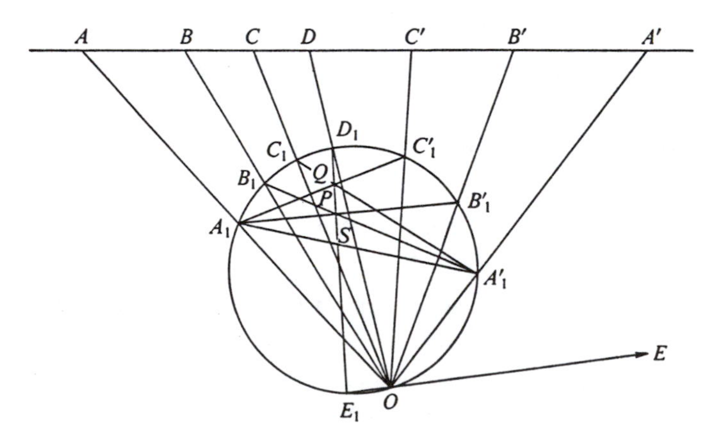

Figure 3.4a

A'_1, B'_1, C'_1 respectively. Let $A_1B'_1$ and A'_1B_1 intersect in P, and $A_1C'_1$ and A'_1C_1 in Q. Draw line PQ. If this line cuts the circle in D_1 and E_1, then the lines OD_1 and OE_1 will intersect the common base of the homographic ranges in their double points D and E.

To prove the above construction, suppose PQ does cut the circle in points D_1 and E_1. Let S be the intersection of PQ and $A_1A'_1$. Then

$$\begin{aligned}(AB,CD) &= O(A_1B_1,C_1D_1) = A'_1(A_1B_1,C_1D_1) = (SP,QD_1)\\ &= A_1(A'_1B'_1,C'_1D_1) = O(A'_1B'_1,C'_1D_1) = (A'B',C'D),\end{aligned}$$

and D is a common, or double, point of the two coaxial homographic ranges. In a similar way it can be shown that E is also a double point of the two ranges.

Conversely, suppose D is a double point of the two coaxial homographic ranges. Let D_1 be the point where OD cuts the circle again, and let A'_1D_1 and A_1D_1 cut the line PQ in M and N respectively. Then

$$(AB,CD) = O(A_1B_1,C_1D_1) = A'_1(A_1B_1,C_1D_1) = (SP,QM).$$

Similarly, $(A'B',C'D) = (SP,QN)$. But, since D is a double point, $(AB,CD) = (A'B',C'D)$. Therefore $(SP,QM) = (SP,QN)$, or $M = N = D_1$, and PQ passes through the point D_1. It follows that our construction gives all the double points of the two distinct coaxial homographic ranges.

3.4.2 REMARKS. (1) Since PQ may intersect, touch, or fail to intersect the circle, two distinct coaxial homographic ranges have two, one, or no double points.

(2) The above construction also solves two other important problems: that of finding the common, or double, rays of two distinct copunctual homographic pencils, and that of finding the common, or double, points of two distinct concyclic homographic ranges.

(3) Note that PQ is the Pascal line of the hexagon $A_1B'_1C_1A'_1B_1C'_1$, whence $B_1C'_1$ and B'_1C_1 also intersect on it.

We now illustrate our new maneuver in a solution of the following problem.

3.4.3 PROBLEM. *Construct a triangle inscribed in one given triangle and circumscribed about another given triangle.*

Referring to Figure 3.4b, let ABC and $A'B'C'$ be the two given triangles. We wish to construct a triangle PQR whose vertices P, Q, R lie on the sides BC, CA, AB of triangle ABC and whose sides QR, RP, PQ pass through the vertices A', B', C' of triangle $A'B'C'$. Take an arbitrary point P_i on BC and draw P_iC' to cut AC in Q_i; next draw Q_iA' to cut AB in R_i; then draw R_iB' to cut BC in P'_i. It is easily seen that range P_i is homographic to range P'_i, and the desired point P is a double point of

Figure 3.4b

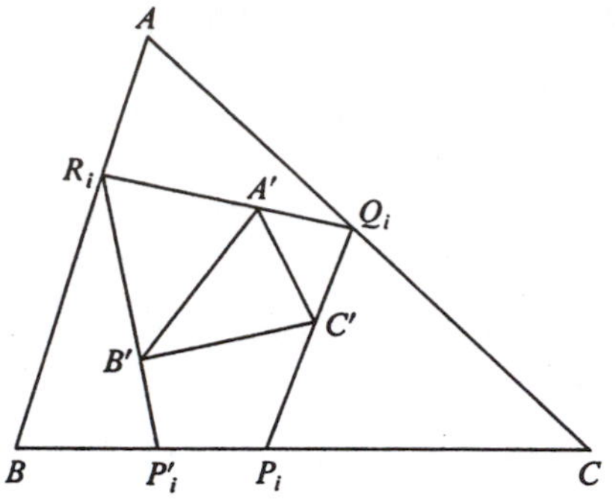

these two coaxial homographic ranges. We may find P by applying Problem 3.4.1.

We now see why the new method is sometimes referred to as the *method of trial and error*; from three *guesses* of the position of a point we find its actual position.

PROBLEMS

1. Through a given point P draw a line intersecting two given lines m *and* m' in corresponding points of two homographic ranges lying on m and m'.
2. Given two homographic pencils, find the pairs of corresponding rays that intersect on a given line m.
3. Given two homographic pencils, find a pair of corresponding rays that intersect at a given angle.
4. Given two homographic pencils, find a pair of corresponding rays that are parallel.
5. Construct a line segment whose extremities lie one each on two given intersecting lines, and which subtends a given angle at each of two given points.
6. Find two points on a given line which are isogonal conjugates with respect to a given triangle.
7. In a given triangle ABC inscribe another, $A'B'C'$, which shall have its sides parallel to three given lines. (This is Problem 15, Section 3.3.)
8. Given four coplanar lines, p, q, r, s, find points A and B on p and q such that the projections of AB on r and s shall have given lengths.
9. Through a given point draw two lines that cut off segments of given lengths on two given lines.

10. Two fixed points O and O' on two fixed lines m and m' are given. Through a fixed point P draw a line cutting m and m' in points A and A' such that $OA/O'A'$ is a given constant. (This problem was the subject matter of Apollonius' treatise *On Proportional Section.*)

11. Two fixed points O and O' on two fixed lines m and m' are given. Through a fixed point P draw a line cutting m and m' in points A and A' such that $(OA)(O'A')$ is a given constant. (This problem was the subject matter of Apollonius' treatise *On Spatial Section.*)

12. Through a given point draw a line to include with two given intersecting lines a triangle of given area.

13. Inscribe a triangle in a given triangle such that its sides will subtend given angles at given points.

14. Solve the *Castillon-Cramer Problem*: Inscribe in a given circle a triangle whose sides, produced if necessary, pass through three given points.

15. Circumscribe a triangle about a given circle such that each vertex shall lie on a given line.

3.5 THE MOHR-MASCHERONI CONSTRUCTION THEOREM

The eighteenth-century Italian geometer and poet, Lorenzo Mascheroni (1750–1800), made the surprising discovery that all Euclidean constructions, insofar as the given and required elements are points, can be made with the compass alone, and that the straightedge is thus a redundant tool. Of course, straight lines cannot be drawn with the compass, but any straight line arrived at in a Euclidean construction can be determined by the compass by finding two points on the line. This discovery appeared in 1797 in Mascheroni's *Geometria del compasso*. Generally speaking, Mascheroni established his results by using the idea of reflection in a line. In 1890, the Viennese geometer, August Adler (1863–1923), published a new proof of Mascheroni's results, using the inversion transformation.

Then an unexpected thing happened. Shortly before 1928, a student of the Danish mathematician Johannes Hjelmslev (1873–1950), while browsing in a bookstore in Copenhagen, came across a copy of an old book, *Euclides Danicus*, published in 1672 by an obscure writer named Georg Mohr. Upon examining the book, Hjelmslev was surprised to find that it contained Mascheroni's discovery, with a different solution, arrived at a hundred and twenty-five years before Mascheroni's publication had appeared.

The present section will be devoted to a proof of the Mohr-Mascheroni discovery. We shall employ the Mascheroni approach, and shall relegate Adler's approach to the problems at the end of the section. We first introduce a compact and elegant way of describing any given construction. The method will become clear from an example, and we choose the construction appearing in Theorem 3.1.2. That construction can be condensed into the following table:

$A(O)$, $O(A)$	$D(B)$, $E(B)$	$O(F)$
D, E	F	

The first line of the table tells us what "circles" we are to draw, and the second line labels the points of intersection so obtained. The table is divided vertically into steps. Reading the above table we have: Step 1. Draw circles $A(O)$ and $O(A)$ to intersect in points D and E. Step 2. Draw circles $D(B)$ and $E(B)$ to intersect in point F. Step 3. Draw circle $O(F)$. It will be noted that this is precisely the construction appearing in the proof of Theorem 3.1.2.

We are now ready to proceed.

3.5.1 PROBLEM. *Given points A, B, C, D, construct, with a modern compass alone, the points of intersection of circle $C(D)$ and line AB.*

Figure 3.5a

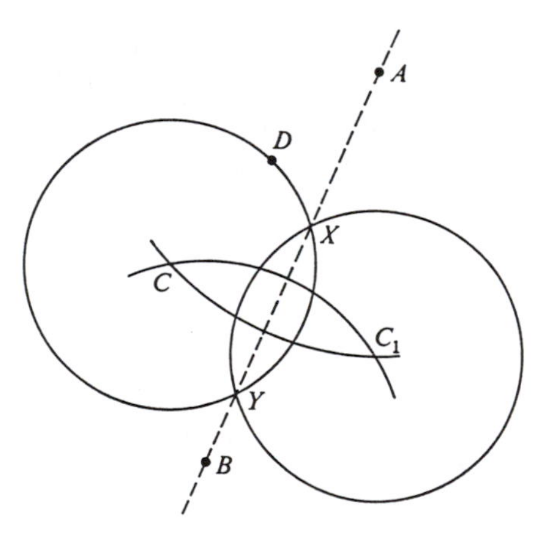

Case 1. C not on AB (see Figure 3.5a).

$A(C)$, $B(C)$	$C(D)$, $C_1(CD)$
C_1	X, Y

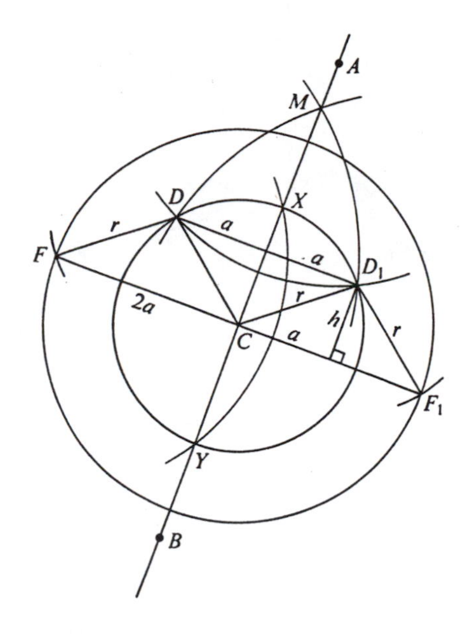

Figure 3.5b

Case 2. C on AB (see Figure 3.5b).

$A(D)$, $C(D)$	$C(DD_1)$, $D(C)$	$C(DD_1)$, $D_1(C)$	$F(D_1)$, $F_1(D)$	$F(CM)$, $C(D)$
D_1	F, 4th vertex of $\square\ CD_1DF$	F_1, 4th vertex of $\square\ CDD_1F_1$	M	X, Y

The proof of case 1 is easy. In case 2, observe (see Figure 3.5b) that

$$\begin{aligned}(CM)^2 &= (FM)^2 - 4a^2 = (FD_1)^2 - 4a^2 = (9a^2 + h^2) - 4a^2 \\ &= 9a^2 + r^2 - a^2 - 4a^2 = 4a^2 + r^2 = (FX)^2.\end{aligned}$$

3.5.2 PROBLEM. *Given points A, B, C, D, construct, with a modern compass alone, the points of intersection of the lines AB and CD (see Figure 3.5c).*

The proof of the construction, given on the next page, is easy. We have C_1D_1GE similar to C_1XCF. But $GD_1 = GE$. Therefore $CX = CF$.

Figure 3.5c

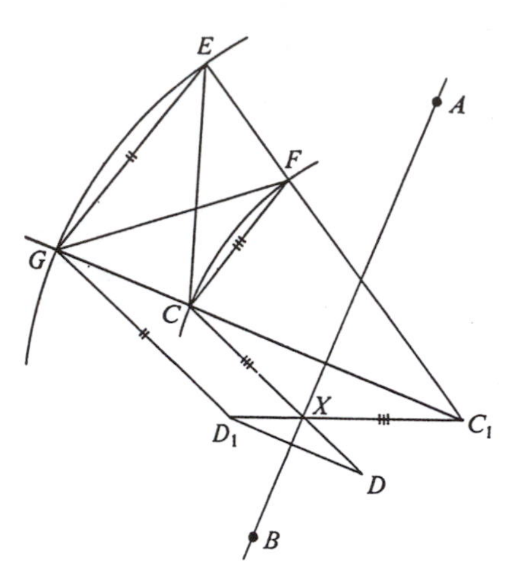

$A(C)$, $B(C)$	$A(D)$, $B(D)$	$C(DD_1)$, $D_1(CD)$	$C_1(G)$, $G(D_1)$	$C_1(C)$, $G(CE)$	$C(F)$, $C_1(CF)$
C_1	D_1	G, collinear with C, C_1	E, either intersection	F, collinear with C_1, E	X

3.5.3 THE MOHR-MASCHERONI CONSTRUCTION THEOREM. *Any Euclidean construction, insofar as the given and required elements are points, may be accomplished with the Euclidean compass alone.*

For in a Euclidean construction, every new point is determined as an intersection of two circles, of a straight line and a circle, or of two straight lines, and the construction, no matter how complicated, is a succession of a finite number of these processes. Because of the equivalence of modern and Euclidean compasses (see Theorem 3.1.2), it is then sufficient to show that with a modern compass alone we are able to solve the following three problems:

I. Given A, B, C, D, find the points of intersection of $A(B)$ and $C(D)$.
II. Given A, B, C, D, find the points of intersection of AB and $C(D)$.
III. Given A, B, C, D, find the point of intersection of AB and CD.

But I is obvious, and we have solved II and III in Problem 3.5.1 and Problem 3.5.2, respectively.

It is to be noted that our proof of the Mohr-Mascheroni construction theorem is more than a mere existence proof, for not only have we shown the existence of a construction using only the Euclidean compass which can replace any given Euclidean construction, but we have shown how such a construction can actually be obtained from the given Euclidean construction. It must be confessed, though, that the resulting construction using only the Euclidean compass would, in all likelihood, be far more complicated than is necessary. The task of finding a "simplest" construction employing only the Euclidean compass, or even only the modern compass, is usually very difficult indeed, and requires considerable ingenuity on the part of the solver.

PROBLEMS

1. Solve or establish, as the case may be, the following constructions using only the *Euclidean* compass:

(a) Given points A and B, find point C on AB produced such that $AB = BC$.

$B(A), A(B)$	$B(A), R(B)$	$B(A), S(B)$
R	S	C

(b) Given points A and B, and a positive integer n, find point C on AB produced such that $AC = n(AB)$.

(c) Given points O, A, B, find the reflection of O in line AB.

(d) Given points O, D, M, find the inverse M' of M in circle $O(D)$ for the following cases.

Case 1. $OM > (OD)/2$.

$O(D), M(O)$	$A(O), B(O)$
A, B	M'

Case 2. $OM \leqq (OD)/2$.

(e) Given noncollinear points A, B, O, find the center Q of the inverse of line AB in circle $O(D)$. Find P, the reflection of O in AB, and then Q, the inverse of P in $O(D)$.

(f) Given points O, D and a circle k not through O, find the center M' of the inverse k' of k in circle $O(D)$. Find M, the inverse of O in k, and then M', the inverse of M in $O(D)$.

(g) Given points A, B, C, D, construct, with a Euclidean compass alone, the points X, Y of intersection of circle $C(D)$ and line AB.

(h) Given points A, B, C, D, construct, with a Euclidean compass alone, the point X of intersection of lines AB and CD.

The above steps essentially constitute Adler's proof of the Mohr-Mascheroni construction theorem. Note that Mascheroni's approach exploits reflections in lines whereas Adler's approach exploits the inversion transformation.

2. Establish the following solution, using a Euclidean compass alone, of the problem of finding the center D of the circle through three given noncollinear points A, B, C.

   Draw circle $A(B)$. Find (by Problem 1, (d)) the inverse C' of C in $A(B)$. Find (by Problem 1, (c)) the reflection D' of A in BC'. Find (by Problem 1, (d)) the inverse D of D' in $A(B)$.

3. Given points A and B, find, with a Euclidean compass alone, the midpoint M of segment AB.

4. On page 268 of Cajori's *A History of Mathematics* we read: "Napoleon proposed to the French mathematicians the problem, to divide the circumference of a circle into four equal parts by the compasses only. Mascheroni does this by applying the radius three times to the circumference; he obtains the arcs AB, BC, CD; then AD is a diameter; the rest is obvious." Complete the "obvious" part of the construction.

5. Look up the following constructions with compass alone which have appeared in *The American Mathematical Monthly*: Problem 3000, Apr. 1924; Problem 3327, June-July 1929; Problem 3706, June-July 1936; Problem E 100, Jan. 1935; Problem E 567, Jan. 1944.

3.6 THE PONCELET-STEINER CONSTRUCTION THEOREM

Though all Euclidean constructions, insofar as the given and required elements are points, are possible with a Euclidean compass alone, it is an easy matter to assure ourselves that not all Euclidean constructions are similarly possible with a Euclidean straightedge alone. To see this, consider the problem of finding the point M midway between two given points A and B, and suppose the problem can be solved with straightedge alone. That is, suppose there exists a finite sequence of lines, drawn according to the restrictions of Euclid's first two postulates, that finally leads from the two given points A and B to the desired midpoint M. Choose a point O outside the plane of construction, and from O project the entire construction upon a second plane not through O. Points A, B, M of the first plane project into points A', B', M' of the second plane, and the sequence of lines leading to the point M in the first plane projects into a sequence of lines leading to the point M' in the second plane. The description of the straightedge construction in the second plane, utilizing the projected sequence of lines, of the point M' from the points A' and B' is exactly like the description of the straightedge construction in the first plane, utilizing the original sequence of lines, of the point M from the points A and B. Since M is the midpoint of AB, it follows, then, that M' must be the midpoint of $A'B'$. But this is absurd, for the midpoint of a line segment need not project into the midpoint of the projected segment. It follows that the simple Euclidean problem of finding the point M midway between two given points A and B is not possible with the straightedge alone.

Our inability to solve all Euclidean constructions with a straightedge alone shows that the straightedge must be assisted with the compass, or with some other tool. It is natural to wonder if the compass can be replaced by some kind of compass less powerful than the Euclidean and modern compasses. As early as the tenth century, the Arab mathematician, Abû'l-Wefâ (940–998), considered constructions carried out with a straightedge and a so-called *rusty compass*, or a compass of fixed opening. Constructions of this sort appeared in Europe in the late fifteenth and early sixteenth centuries and engaged the attention, among others, of the great artists Albrecht Dürer (1471–1528) and Leonardo da Vinci (1452–1519). In this early work, the motivation was a practical one, and the radius of the rusty compass was chosen as some length convenient for the problem at hand. In the middle of the sixteenth century, a new viewpoint on the matter was adopted by Italian mathematicians. The motivation became a purely academic one, and the radius of the rusty compass was considered as arbitrarily assigned at the start. A number of writers showed how all the constructions in Euclid's *Elements* can be carried out with a straightedge and a given rusty compass. Some real ingenuity is required to accomplish this, as becomes evident to anyone who tries to construct with a straightedge and a rusty compass the triangle whose three sides are given.

But the fact that all the constructions in Euclid's *Elements* can be carried out with a straightedge and a given rusty compass does not prove that a straightedge and a rusty compass are together equivalent to a straightedge and a Euclidean compass. This equivalence was first indicated in 1822 by Victor Poncelet, who stated, with a suggested method of proof, that all Euclidean constructions can be carried out with the straightedge alone in the presence of a single circle and its center. This implies that all Euclidean constructions can be carried out with a straightedge and a rusty compass, and that, moreover, the rusty compass need be used *only once*, and thenceforth discarded. In 1833, Jacob Steiner gave a complete and systematic treatment of Poncelet's theorem. It is our aim in this section to develop a proof of the above Poncelet-Steiner theorem.

3.6.1 PROBLEM. *Given points A, B, U, P, where U is the midpoint of AB and P is not on line AB, construct, with straightedge alone, the line through P parallel to line AB.*

Draw lines AP, BP, and an arbitrary line BC through B cutting AP in

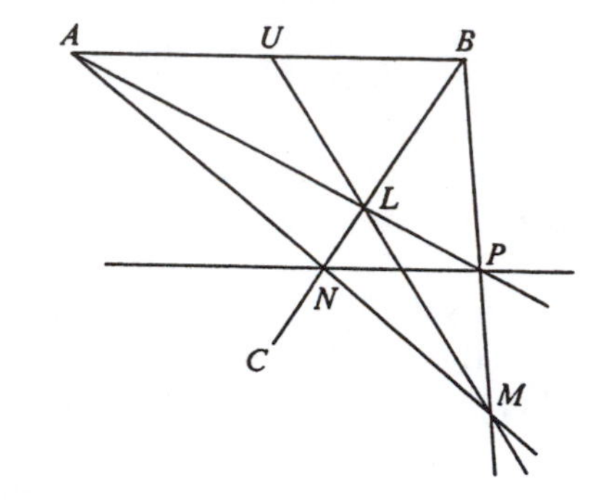

Figure 3.6a

L (see Figure 3.6a). Draw UL to cut BP in M. Draw AM to cut BC in N. Then PN is the sought parallel. The proof follows from the fact that in the quadrangle $ABPN$, PN must cut AB in the harmonic conjugate of U for A and B; since U is the midpoint of AB, the harmonic conjugate is the point at infinity on AB.

3.6.2 PROBLEM. *Given a circle k with its center O, a line AB, and a point P not on AB, construct, with straightedge alone, the line through P parallel to line AB.*

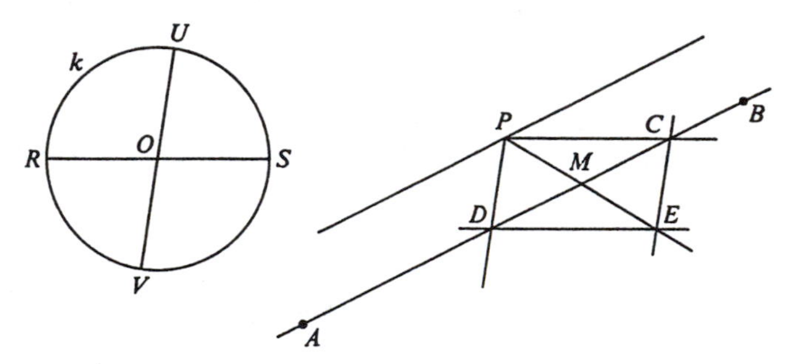

Figure 3.6b

Referring to Figure 3.6b, draw any two diameters RS, UV of k not parallel to AB. Draw (by Problem 3.6.1) PC parallel to RS to cut AB in C, PD parallel to UV to cut AB in D, DE and CE parallel to RS and UV respectively to cut in E. Let PE cut DC in M. We now have a bisected segment on line AB and we may proceed as in Problem 3.6.1.

3.6.3 PROBLEM. *Given a circle k with its center O, a line AB, and a point P not on AB, construct, with straightedge alone, the reflection of P in AB.*

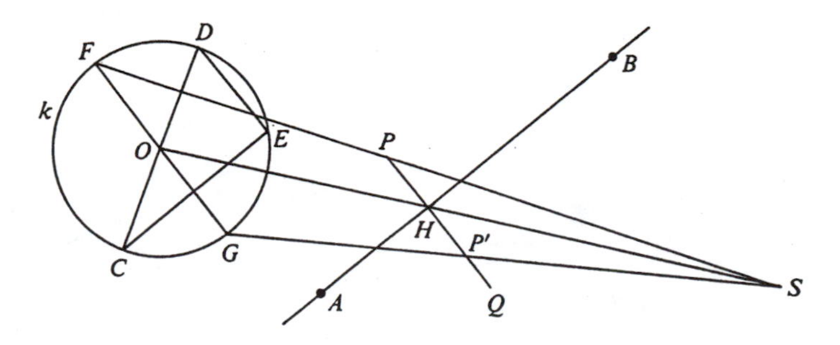

Figure 3.6c

Referring to Figure 3.6c, draw any diameter CD of k not parallel to AB. Draw (by Problem 3.6.2) chord CE parallel to AB, and then diameter FOG and line PQ parallel to DE. Let PQ cut AB in H and let FP and OH meet in S. Then GS cuts PQ in the sought point P'. If FP and OH are parallel, draw GS through G parallel to FP to cut PQ in the sought point P'. In either case, the proof follows from the fact that PHP' is homothetic to FOG.

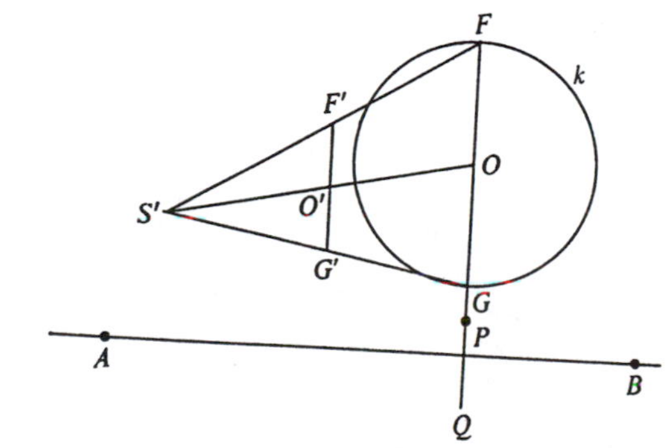

Figure 3.6d

If PQ collines with FG (see Figure 3.6d) the above construction fails. In this case connect F, O, G with any point S' not on FG and cut these joins by a line parallel to FOG, obtaining the points F', O', G'. We may now proceed as before.

3.6.4 PROBLEM. *Given a circle k with its center O, and points A, B, C, D, construct, with straightedge alone, the points of intersection of line AB and circle $C(D)$.*

Referring to Figure 3.6e, draw radius OR parallel to CD and let OC, RD meet in S. Draw any line DL and then RM parallel to DL to cut SL in N. Draw NP parallel to AB to cut k in U and V. Now draw US, VS to cut AB in the sought points X and Y.

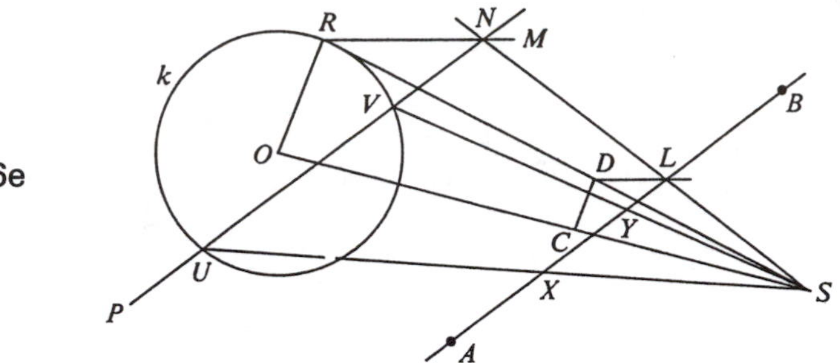

Figure 3.6e

If O, D, C colline, find (by Problem 3.6.3) the reflection D' of D in any line through C, and proceed as before.

If $CD = OR$, then S is at infinity, but a construction can be carried out similar to the above by means of parallels.

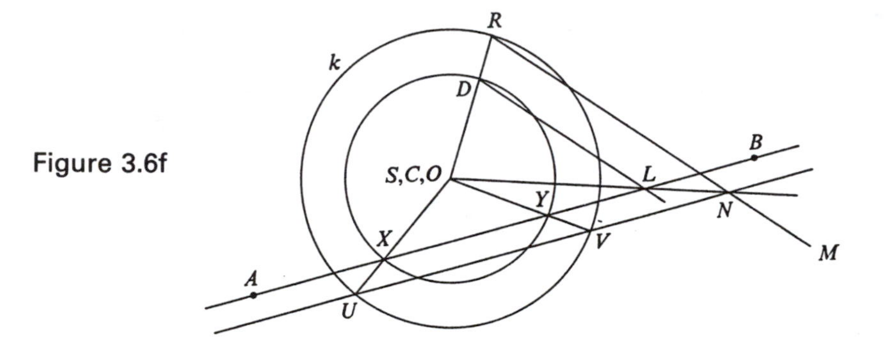

Figure 3.6f

If $C = O$, take $S = O$ (see Figure 3.6f).

In all cases, a proof is easy by simple homothety considerations.

3.6.5 Problem. *Given a circle k with its center O, and points A, B, C, D, construct, with straightedge alone, the radical axis of circles A(B) and C(D).*

Referring to Figure 3.6g, draw $\overline{AQ}$ parallel to $\overline{CD}$. Find (by Problem 3.6.4) the point P where AQ cuts $A(B)$. AC and PD determine the external center of similitude S of $A(B)$ and $C(D)$. Draw any two lines through S to cut the circles in E, F, G, H and J, K, L, M, which can be found by Problem 3.6.4. Let JF and MG intersect in R and EK and HL in T. Then RT is the sought radical axis.

A proof may be given as follows. Since arcs FK and HM have the same angular measure, $\sphericalangle HEK = \sphericalangle HLM$. Therefore $HEKL$ is concyclic,

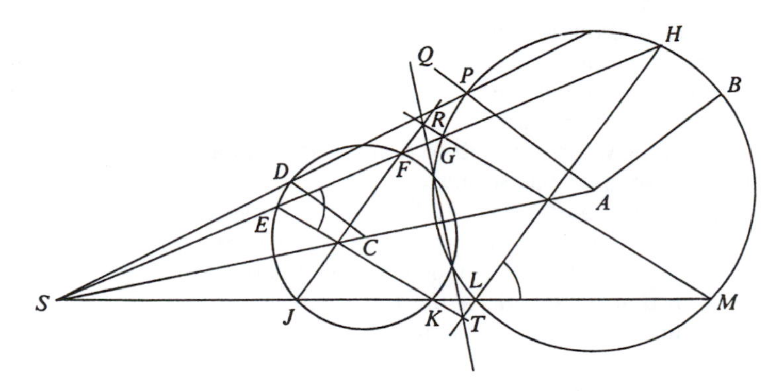

Figure 3.6g

whence $(TK)(TE) = (TL)(TH)$, and T is on the radical axis of $A(B)$ and $C(D)$. R is similarly on the radical axis of $A(B)$ and $C(D)$.

3.6.6 PROBLEM. *Given a circle k with its center O, and points A, B, C, D, construct, with straightedge alone, the points of intersection of circles A(B) and C(D).*

Find, by Problem 3.6.5, the radical axis of $A(B)$ and $C(D)$. Then, by Problem 3.6.4, find the points of intersection of this radical axis with $A(B)$.

3.6.7 THE PONCELET-STEINER CONSTRUCTION THEOREM. *Any Euclidean construction, insofar as the given and required elements are points, may be accomplished with straightedge alone in the presence of a given circle and its center.*

We leave the proof, which can be patterned after that of Theorem 3.5.3, to the reader.

PROBLEMS

1. It has been shown* that the Georg Mohr mentioned in Section 3.5 was the author of an anonymously published booklet entitled *Compendium Euclidis Curiosi*, which appeared in 1673 and which in effect shows that all the constructions of Euclid's *Elements* are possible with a straightedge and a given rusty compass. Solve, with a straightedge and a given rusty compass, the following first fourteen constructions found in Mohr's work (see the second reference in the footnote).

* See A. E. Hallerberg, "The geometry of the fixed-compass," *The Mathematics Teacher*, Apr. 1959, pp. 230–244, and A. E. Hallerberg, "Georg Mohr and Euclidis Curiosi," *The Mathematics Teacher*, Feb. 1960, pp. 127–132.

(1) To divide a given line into two equal parts.
(2) To erect a perpendicular to a line from a given point in the given line.
(3) To construct an equilateral triangle on a given side.
(4) To erect a perpendicular to a line from a given point off the given line.
(5) Through a given point to draw a line parallel to a given line.
(6) To add two given line segments.
(7) To subtract a shorter segment from a given segment.
(8) Upon the end of a given line to place a given segment perpendicularly.
(9) To divide a line into any number of equal parts.
(10) Given two lines, to find the third proportional.
(11) Given three lines, to find the fourth proportional.
(12) To find the mean proportional to two given segments.
(13) To change a given rectangle into a square.
(14) To draw a triangle, given the three sides.

2. With a straightedge alone find the polar p of a given point P for a given circle k.

3. (a) With a straightedge alone construct the tangents to a given circle k from a given external point P.
(b) With a straightedge alone construct the tangent to a given circle k at a given point P on k.

4. Given three collinear points A, B, C with $AB = BC$, construct, with a straightedge alone, (a) the point X on line AB such that $AX = n(AB)$, where n is a given positive integer, (b) the point Y on AB such that $AY = (AB)/n$.

5. Supply the proof of Theorem 3.6.7.

6. Look up the following constructions with straightedge alone that have appeared in *The American Mathematical Monthly*: Problem 3089, Jan. 1929; Problem 3137, June-July 1926; Problem E 539, June-July 1943 and Aug.-Sept. 1948; Problem E 793, Nov. 1948.

3.7 SOME OTHER RESULTS (OPTIONAL)

There are other construction theorems that are of great interest to devotees of the construction game, but space forbids us from doing much more than briefly to comment on a few of them.

It will be recalled that the Poncelet-Steiner theorem states that any Euclidean construction can be carried out with a straightedge alone in the presence of a circle *and its center*. It can be shown that the center of a given circle cannot be found by straightedge alone. (The method is similar to that employed at the start of Section 3.6 to show that the point midway between two given points cannot be found with straightedge alone.) It follows that, in the Poncelet-Steiner theorem, the center of the given circle is a necessary piece of data. But the interesting question arises: How many circles in the plane do we need in order to find their centers with a straightedge alone? It has been shown that two circles suffice provided they intersect, are tangent, or are concentric; otherwise three noncoaxial circles are necessary. It also can be shown that the centers of any two given circles can be found with straightedge alone if we are given either a center of similitude of the two circles, or a point on their radical axis, and if only one circle is given, its center can be found with straightedge alone if we are also given a parallelogram somewhere in the plane of construction. Perhaps the most remarkable finding in connection with the Poncelet-Steiner theorem is that not all of the given circle is needed, but that *all Euclidean constructions are solvable with straightedge alone in the presence of a circular arc, no matter how small, and its center.*

Adler and others have shown that *all Euclidean constructions are solvable with a double-edged ruler, be the two edges parallel or intersecting at an angle.* Examples of the latter type of two-edged ruler are a carpenter's square and a draughtsman's triangle. It is interesting that while an increase in the number of compasses will not enable us to solve anything more than the Euclidean constructions, two carpenter squares make it possible for us to duplicate a given cube and to trisect an arbitrary angle. These latter problems are also solvable with compass and a *marked* straightedge—that is, a straightedge bearing two marks along its edge. Various tools and linkages have been invented that will solve certain problems beyond those solvable with Euclidean tools alone. Another interesting discovery is that *all Euclidean constructions, insofar as the given and required elements are points, can be solved without any tools whatever, by simply folding and creasing the paper representing the plane of construction.*

Suppose we consider a point of intersection of two loci as *ill-defined* if the two loci intersect at the point in an angle less than some given small angle θ, and that otherwise the point of intersection will be considered as *well-defined.* The question arises: Can any given Euclidean construction be accomplished with the Euclidean tools by using only well-defined intersections? Interestingly enough, the answer is an affirmative one, and the result can be strengthened. In addition to the small angle θ, let us be given a small linear distance d, and suppose we define a Euclidean construction to be *well-defined* if it utilizes only well-defined intersections, straight lines determined by pairs of points which are a greater distance than d apart, and circles with radii greater than d. Then it can be shown that *any Euclidean construction can be accomplished by a well-defined one.* Another, and easy to prove, theorem is the following concerning constructions in a limited space: *Let a Euclidean construction be composed of given, auxiliary, and required loci G, A, R respectively, and let S be a convex region placed on the construction such that S contains at least a part G', A', R' of each locus of G, A, R. Then there exists a Euclidean construction determining R', insofar as each locus of R' is considered determined by points, from the loci of G', the construction being performed entirely inside S.*

Euclid's first postulate presupposes that our straightedge is as long as we wish, so that we can draw the line determined by *any* two given points A and B. Suppose we have a straightedge of small finite length ε, such that it will not span the distance between A and B. Can we, with our little ε-straightedge alone, still draw the line joining A and B? This would surely seem to be impossible but, curiously enough, after a certain amount of trial and error, the join can be drawn. Let us show how this may be accomplished. After a finite amount of trial and error it is always possible

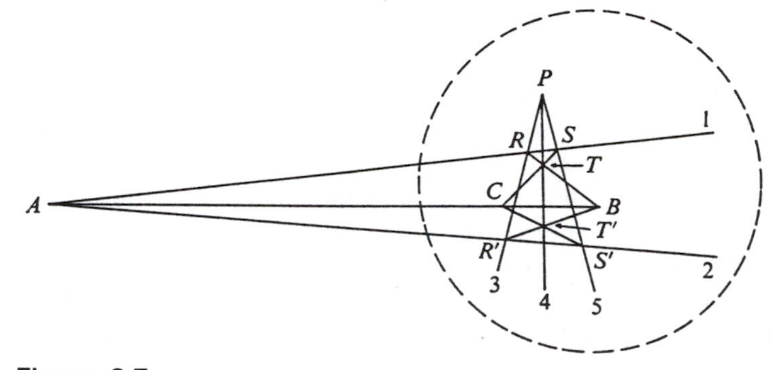

Figure 3.7a

(see Figure 3.7a) to obtain, with our ε-straightedge, two lines 1 and 2 through A and lying so close to B that the ε-straightedge will comfortably span lines 1, 2, and AB. Choose a point P sufficiently close to B and draw three lines 3, 4, 5 through P. Let line 3 cut lines 1 and 2 in points R and R', and let line 5 cut lines 1 and 2 in points S and S'. Draw BR and BR' to cut line 4 in T and T'. Finally, draw ST and $S'T'$ to intersect in C. Then, since triangles RST, $R'S'T'$ are copolar at P, they are (by Desargues' Theorem) coaxial, and points A, B, C must be collinear. But our ε-straightedge will span B and C, and thus the line BA can be drawn.

The problem of finding the "best" Euclidean solution to a required construction has also been considered, and a science of *geometrography* was developed in 1907 by Émile Lemoine for quantitatively comparing one construction with another. To this end, Lemoine considered the following five operations:

S_1: to make the straightedge pass through one given point,
S_2: to rule a straight line,
C_1: to make one compass leg coincide with a given point,
C_2: to make one compass leg coincide with an arbitrary point of a given locus,
C_3: to describe a circle.

If the above operations are performed m_1, m_2, n_1, n_2, n_3 times in a construction, then $m_1S_1 + m_2S_2 + n_1C_1 + n_2C_2 + n_3C_3$ is regarded as the *symbol* of the construction. The total number of operations, $m_1 + m_2 + n_1 + n_2 + n_3$, is called the *simplicity* of the construction, and the total number of coincidences, $m_1 + n_1 + n_2$, is called the *exactitude* of the construction. The total number of loci drawn is $m_2 + n_3$, the difference between the simplicity and the exactitude of the construction.

As simple examples, the symbol for drawing the straight line through A and B is $2S_1 + S_2$, and that for drawing the circle with center C and radius AB is $3C_1 + C_3$.

Of course the construction game has been extended to three-dimensional space, but we shall not consider this extension here.

PROBLEMS

1. Prove that the center of a given circle cannot be found with straightedge alone.

2. (a) Prove that, given a parallelogram somewhere in the plane of construction, we can, with straightedge alone, draw a line parallel to any given line AB through any given point P.
   (b) Prove that, given a parallelogram somewhere in the plane of construction, we can, with straightedge alone, find the center of any given circle k.

3. (a) Prove that, with straightedge alone, we can find the common center of two given concentric circles.
   (b) Prove that, with straightedge alone, we can find the centers of two given tangent circles.
   (c) Prove that, with straightedge alone, we can find the centers of two given intersecting circles.

4. Prove that the centers of any two given circles can be found with straightedge alone if we are given a center of similitude of the two circles.

5. Use a ruler having two parallel edges to solve the following problems:
   (a) To obtain a bisected segment on a given line AB.
   (b) Through a given point P to draw a line parallel to a given line AB.
   (c) Through a given point P to draw a line perpendicular to a given line AB.
   (d) Given points A, B, C, D, to find the points of intersection of line AB and circle $C(D)$.
   (e) Given points A, B, C, D, to find the points of intersection of circles $A(B)$ and $C(D)$.

6. Show that Problem 5 establishes the theorem: *All Euclidean constructions, insofar as the given and required elements are points, can be accomplished with a ruler having two parallel edges.*

7. Look up Euclidean constructions by paper folding in R. C. Yates, *Geometrical Tools*, Educational Publishers, St. Louis (revised 1949).

8. Look up well-defined Euclidean constructions in Howard Eves and Vern

Hoggatt, "Euclidean constructions with well-defined intersections," *The Mathematics Teacher*, vol. 44, no. 4 (April 1951), pp. 262–63.

9. Establish the theorem of Section 3.7 concerning Euclidean constructions in a limited space.

10. Let us be given two curves m and n, and a point O. Suppose we permit ourselves to mark, on a given straightedge, a segment MN, and then to adjust the straightedge so that it passes through O and cuts the curves m and n with M on m and N on n. The line drawn along the straightedge is then said to have been drawn by "the insertion principle." Problems beyond the Euclidean tools can often be solved with these tools if we also permit ourselves to use the insertion principle. Establish the correctness of the following constructions, each of which uses the insertion principle.
(a) Let AB be a given segment. Draw $\measuredangle ABM = 90°$ and $\measuredangle ABN = 120°$. Now draw ACD cutting BM in C and BN in D and such that $CD = AB$. Then $(AC)^3 = 2(AB)^3$. Essentially this construction for duplicating a cube was given in publications by Viète (1646) and Newton (1728).
(b) Let AOB be any central angle in a given circle. Through B draw a line BCD cutting the circle again in C, AO produced in D, and such that $CD = OA$, the radius of the circle. Then $\measuredangle ADB = (\measuredangle AOB)/3$. This solution of the trisection problem is implied by a theorem given by Archimedes (*ca.* 240 B.C.).

11. Over the years many mechanical contrivances, linkage machines, and compound compasses, have been devised to solve the trisection problem. An interesting and elementary implement of this kind is the so-called *tomahawk*. The inventor of the tomahawk is not known, but the instrument was described in a book in 1835. To construct a tomahawk, start with a line segment RU

Figure 3.7b

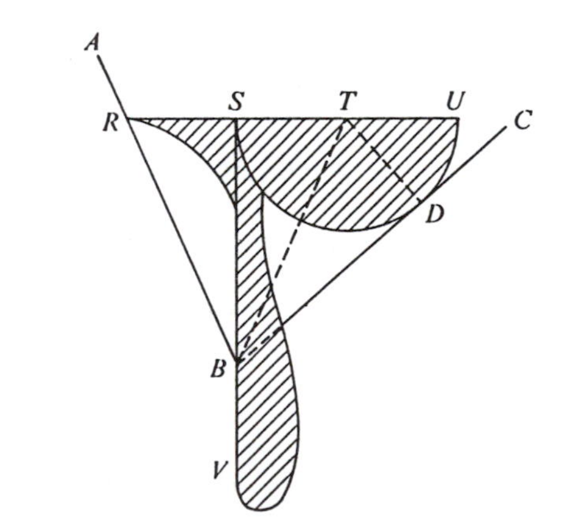

trisected at S and T (see Figure 3.7b). Draw a semicircle on SU as diameter and draw SV perpendicular to RU. Complete the instrument as indicated in the figure. To trisect an angle ABC with the tomahawk, place the implement on the angle so that R falls on BA, SV passes through B, and the semicircle touches BC, at D say. Show that BS and BT then trisect the angle.

The tomahawk may be constructed with straightedge and compass on tracing paper and then adjusted on the given angle. By this subterfuge we may trisect an angle with straightedge and compass.

12. A construction using Euclidean tools but requiring an infinite number of operations is called an *asymptotic Euclidean construction*. Establish the following two constructions of this type for solving the trisection and the quadrature problems. (For an asymptotic Euclidean solution of the duplication problem, see T. L. Heath, *History of Greek Mathematics*, vol. 1 [New York: Dover, 1981], pp. 268–270.)

    (a) Let OT_1 be the bisector of $\sphericalangle AOB$, OT_2 that of $\sphericalangle AOT_1$, OT_3 that of $\sphericalangle T_2OT_1$, OT_4 that of $\sphericalangle T_3OT_2$, OT_5 that of $\sphericalangle T_4OT_3$, and so forth. Then $\lim_{i\to\infty} OT_i = OT$, one of the trisectors of $\sphericalangle AOB$. (This construction was given by Fialkowski, 1860.)

    (b) On the segment AB_1 produced mark off $B_1B_2 = AB_1$, $B_2B_3 = 2(B_1B_2)$, $B_3B_4 = 2(B_2B_3)$, and so forth. With $B_1, B_2, B_3, \ldots$ as centers draw the circles $B_1(A), B_2(A), B_3(A), \ldots$. Let M_1 be the midpoint of the semicircle on AB_2. Draw B_2M_1 to cut circle $B_2(A)$ in M_2, B_3M_2 to cut circle $B_3(A)$ in $M_3, \ldots$. Let N_i be the projection of M_i on the common tangent of the circles at A. Then $\lim_{i\to\infty} AN_i =$ quadrant of circle $B_1(A)$.

13. Find the symbol, simplicity, and exactitude for the following familiar constructions of a line through a given point A and parallel to a given line MN.

    (a) Through A draw any line to cut MN in B. With any radius r draw the circle $B(r)$ to cut MB in C and AB in D. Draw circle $A(r)$ to cut AB in E. Draw circle $E(CD)$ to cut circle $A(r)$ in X. Draw AX, obtaining the required parallel.

    (b) With any suitable point D as center draw circle $D(A)$ to cut MN in B and C. Draw circle $C(AB)$ to cut circle $D(A)$ in X. Draw AX.

    (c) With any suitable radius r draw the circle $A(r)$ to cut MN in B. Draw circle $B(r)$ to cut MN in C. Draw circle $C(r)$ to cut circle $A(r)$ in X. Draw AX.

14. The Arabians were interested in constructions on a spherical surface. Consider the following problems, to be solved with Euclidean tools and appropriate planar constructions.

    (a) Given a material sphere, find its diameter.

    (b) On a given sphere locate the vertices of an inscribed cube.

    (c) On a given material sphere locate the vertices of an inscribed regular tetrahedron.

15. J. S. Mackay, in his *Geometrography*, regarded the following construction as probably the simplest way of finding the center of a given circle; the construction is credited to a person named Swale, in about 1830.

    With any point O on the circumference of the given circle as center and any convenient radius, describe an arc PRQ, cutting the circle in P and Q. With Q as center, and with the same radius, describe an arc OR, cutting arc PRQ in R. Draw PR, cutting the circle again in L. Then RL is the radius of

the given circle, and the center K may now be found by means of two intersecting arcs.
(a) Establish the correctness of Swale's construction.
(b) Find the simplicity and exactitude of the construction.

3.8 THE REGULAR SEVENTEEN-SIDED POLYGON (OPTIONAL)

In Book IV of Euclid's *Elements* are found constructions, with straightedge and compass, of regular polygons of 3, 4, 5, 6, and 15 sides. By successive angle, or arc, bisections, we may then with Euclidean tools construct regular polygons having $3(2^n)$, $4(2^n)$, $5(2^n)$, and $15(2^n)$ sides, where $n = 0, 1, \ldots$. Not until almost the nineteenth century was it suspected that any other regular polygons could be constructed with these limited tools.

In 1796, the eminent German mathematician Karl Friedrich Gauss, then only 19 years old, developed the theory that shows that a regular polygon having a *prime* number of sides can be constructed with Euclidean tools if and only if that number is of the form

$$f(n) = 2^{2^n} + 1.$$

For $n = 0, 1, 2, 3, 4$ we find $f(n) = 3, 5, 17, 257, 65537$, all prime numbers. Thus, unknown to the Greeks, regular polygons of 17, 257, and 65537 sides can be constructed with straightedge and compass. For no other value of n, than those listed above, is it known that $f(n)$ is a prime number. Indeed, after considerable effort, $f(n)$ has actually been shown to be composite for a number of values of $n > 4$, and the general feeling among number theorists today is that $f(n)$ is probably composite for all $n > 4$. The numbers $f(n)$ are usually referred to as *Fermat numbers*, because in about 1640 the great French number theorist, Pierre de Fermat (1601–1665), conjectured (incorrectly, as it has turned out) that $f(n)$ is prime for all nonnegative integral n.

Many Euclidean constructions of the regular polygon of 17 sides (the regular heptadecagon) have been given. In 1832, F. J. Richelot published an investigation of the regular polygon of 257 sides, and a Professor Hermes of Lingen gave up ten years of his life to the problem of constructing a regular polygon of 65537 sides. It has been reported that it was Gauss's discovery, at the age of 19, that a regular polygon of 17 sides can be constructed with straightedge and compass that decided him to devote his life to mathematics. His pride in this discovery is evidenced by his request that a regular polygon of 17 sides be engraved on his tombstone. Although the request was never fulfilled, such a polygon is found on a monument to Gauss erected at his birthplace in Braunschweig (Brunswick).

We will not here take up Gauss's general theory of polygonal construction, but will merely give an elementary synthetic treatment of the regular

17-sided polygon. Some of the easy details will be left to the reader, who may find it interesting to apply an analogous treatment to the construction of a regular pentagon. We commence with some preliminary matter.

3.8.1 LEMMA. *If C and D are two points on a semicircumference ACDB of radius R, with C lying between A and D, and if C′ is the reflection of C in the diameter AB, then:* (1) $(AC)(BD) = R(C'D - CD)$, (2) $(AD)(BC) = R(C'D + CD)$, (3) $(AC)(BC) = R(CC')$.

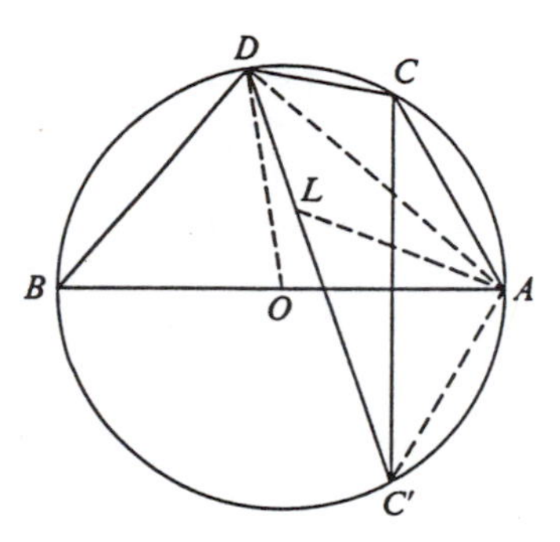

Figure 3.8a₁

(1) On DC' mark off $DL = DC$ and connect A with L, A with D, A with C', and D with the center O of the semicircle (see Figure $3.8a_1$). In triangles DCA and DLA, $DC = DL$, $DA = DA$, and $\sphericalangle CDA = \sphericalangle LDA$. Therefore triangles DCA and DLA are congruent, whence $AL = AC = AC'$, and triangles BOD and $C'AL$ are isosceles. But $\sphericalangle ABD = \sphericalangle AC'D$, whence triangles BOD and $C'AL$ are similar and $BD/C'L = OB/AC'$. That is, $BD/(C'D - CD) = R/AC$, or $(AC)(BD) = R(C'D - CD)$.

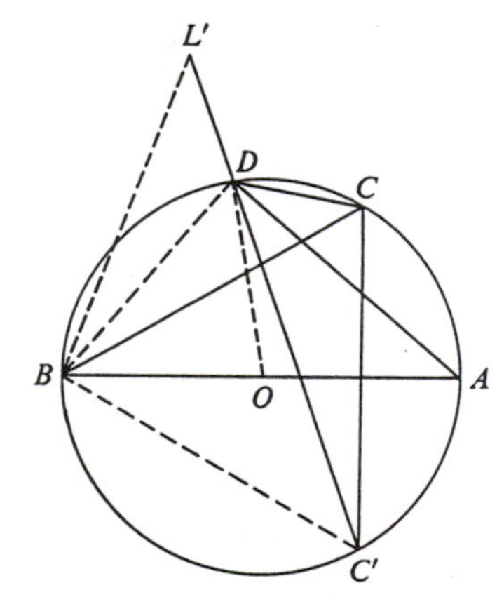

Figure 3.8a₂

(2) On $C'D$ produced, mark off $DL' = DC$ and connect B with L', B with D, B with C', and D with O (see Figure $3.8a_2$). In triangles DCB

and $DL'B$, $DC = DL'$, $DB = DB$, and $\measuredangle CDB = (1/2)(\text{arc } CAC'B) = (1/2)(\text{arc } BDCC') = (1/2)(\text{arc } BD + \text{arc } DCC') = \measuredangle L'DB$. Therefore triangles DCB and $DL'B$ are congruent, whence $BL' = BC = BC'$, and triangles AOD and $C'BL'$ are isosceles. But $\measuredangle BAD = \measuredangle BC'D$, whence triangles AOD and $C'BL'$ are similar and $AD/C'L' = OA/BC'$. That is, $AD/(C'D + CD) = R/BC$, or $(AD)(BC) = R(C'D + CD)$.

(3) Since $\measuredangle BCA = 90°$, we have $(AC)(BC) = 2(\text{area triangle } BCA) = (BA)(CC')/2 = R(CC')$.

3.8.2 THEOREM. *Let the circumference of a circle of radius R be divided into 17 equal parts and let AB be the diameter through one of the points of section A and the midpoint B of the opposite arc. Let the points of section on each side of the diameter AB be consecutively named $C_1, C_2, \ldots, C_8$ and $C'_1, C'_2, \ldots, C'_8$ beginning next to A (see Figure 3.8b).*

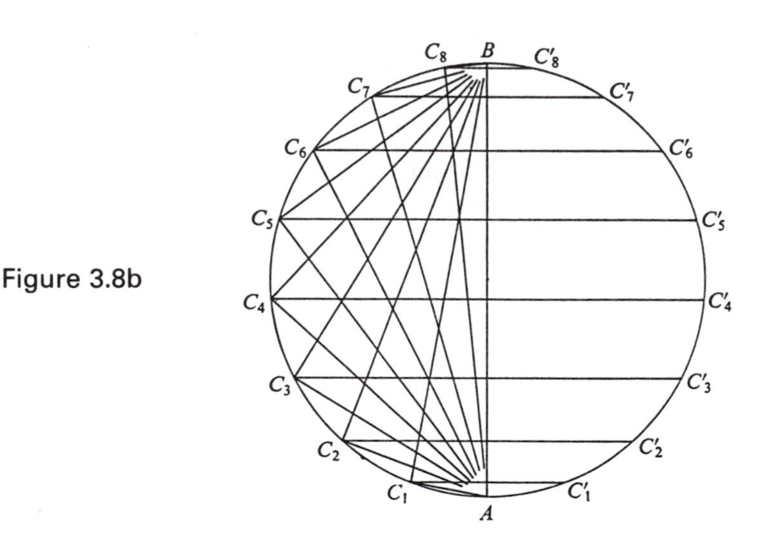

Figure 3.8b

Then

$$(BC_1)(BC_2)(BC_3)(BC_4)(BC_5)(BC_6)(BC_7)(BC_8) = R^8.$$

For, by Lemma 3.8.1 (3), we have $(AC_i)(BC_i) = R(C_iC'_i)$, $i = 1, \ldots, 8$. But, since the circumference is divided into equal parts, $AC_1 = C_8C'_8$, $AC_2 = C_1C'_1$, $AC_3 = C_7C'_7$, $AC_4 = C_2C'_2$, $AC_5 = C_6C'_6$, $AC_6 = C_3C'_3$, $AC_7 = C_5C'_5$, $AC_8 = C_4C'_4$. The theorem now readily follows.

3.8.3 COROLLARY. $(BC_1)(BC_2)(BC_4)(BC_8) = R^4$.

3.8.4 Corollary. $(BC_3)(BC_5)(BC_6)(BC_7) = R^4$.

3.8.5 Theorem. *In the notation of Theorem 3.8.2, we have*

$$(BC_1)(BC_4) = R(BC_3 + BC_5), \qquad (BC_2)(BC_8) = R(BC_6 - BC_7),$$
$$(BC_3)(BC_5) = R(BC_2 + BC_8), \qquad (BC_6)(BC_7) = R(BC_1 - BC_4).$$

Let us denote the midpoints of arcs AC_1, C_1C_2, $C_2C_3, \ldots, C_7C_8$ by $C_{0.5}$, $C_{1.5}$, $C_{2.5}, \ldots, C_{7.5}$, and the midpoints of arcs AC'_1, $C'_1C'_2$, $C'_2C'_3$, $\ldots, C'_7C'_8$ by $C'_{0.5}$, $C'_{1.5}$, $C'_{2.5}, \ldots, C'_{7.5}$.

Now, by Lemma 3.8.1 (2),

$$(AC_{4.5})(BC_1) = R(C'_{4.5}C_1 + C_{4.5}C_1).$$

But $AC_{4.5} = BC_4$, $C'_{4.5}C_1 = BC_3$, $C_{4.5}C_1 = BC_5$, whence, by substitution,

$$(BC_1)(BC_4) = R(BC_3 + BC_5).$$

By Lemma 3.8.1 (1),

$$(AC_{0.5})(BC_2) = R(C'_2C_{0.5} - C_2C_{0.5}).$$

But $AC_{0.5} = BC_8$, $C'_2C_{0.5} = BC_6$, $C_2C_{0.5} = BC_7$, whence, by substitution,

$$(BC_2)(BC_8) = R(BC_6 - BC_7).$$

In like manner, using Lemma 3.8.1 (1) and (2), we can show that

$$(BC_3)(BC_5) = R(BC_2 + BC_8), \qquad (BC_6)(BC_7) = R(BC_1 - BC_4).$$

3.8.6 Notation. Set $M = BC_3 + BC_5$, $N = BC_6 - BC_7$, $P = BC_2 + BC_8$, $Q = BC_1 - BC_4$.

3.8.7 Theorem. $MN = PQ - R^2$.

By Theorem 3.8.5 we have

$$M = BC_3 + BC_5 = (BC_1)(BC_4)/R,$$
$$N = BC_6 - BC_7 = (BC_2)(BC_8)/R,$$

whence, by Corollary 3.8.3,

$$MN = (BC_1)(BC_2)(BC_4)(BC_8)/R^2 = R^4/R^2 = R^2.$$

Similarly, using Corollary 3.8.4, we can show that $PQ = R^2$.

3.8.8 Theorem. $BC_1 - BC_2 + BC_3 - BC_4 + BC_5 - BC_6 + BC_7 - BC_8 = R$.

We have, by Theorem 3.8.7,

$$R^2 = MN = (BC_3 + BC_5)(BC_6 - BC_7)$$
$$= (BC_3)(BC_6) + (BC_5)(BC_6) - (BC_3)(BC_7) - (BC_5)(BC_7).$$

But, by Lemma 3.8.1 (1), we have

$$(BC_3)(BC_6) = R(BC_3 - BC_8), \qquad (BC_5)(BC_6) = R(BC_1 - BC_6),$$
$$(BC_3)(BC_7) = R(BC_4 - BC_7), \qquad (BC_5)(BC_7) = R(BC_2 - BC_5).$$

Substituting, we obtain

$$BC_1 - BC_2 + BC_3 - BC_4 + BC_5 - BC_6 + BC_7 - BC_8 = R.$$

3.8.9 THEOREM. *(1)* $(M - N) - (P - Q) = R$, *(2)* $(M - N)(P - Q) = 4R^2$.

(1) We have (using Theorem 3.8.8)

$$\begin{aligned}(M - N) - (P - Q) \\ &= (BC_3 + BC_5) - (BC_6 - BC_7) - (BC_2 + BC_8) + (BC_1 - BC_4) \\ &= BC_1 - BC_2 + BC_3 - BC_4 + BC_5 - BC_6 + BC_7 - BC_8 = R.\end{aligned}$$

(2) By Theorem 3.8.5,

$$\begin{aligned}(M - N)(P - Q) \\ &= [(BC_3 + BC_5) - (BC_6 - BC_7)][(BC_2 + BC_8) - (BC_1 - BC_4)] \\ &= -(BC_1)(BC_3) - (BC_1)(BC_5) + (BC_1)(BC_6) - (BC_1)(BC_7) \\ &\quad + (BC_2)(BC_3) + (BC_2)(BC_5) - (BC_2)(BC_6) + (BC_2)(BC_7) \\ &\quad + (BC_4)(BC_3) + (BC_4)(BC_5) - (BC_4)(BC_6) + (BC_4)(BC_7) \\ &\quad + (BC_8)(BC_3) + (BC_8)(BC_5) - (BC_8)(BC_6) + (BC_8)(BC_7).\end{aligned}$$

But, by Lemma 3.8.1 (1) and (2), we can show that

$$\begin{aligned}
(BC_1)(BC_3) &= R(BC_2 + BC_4), & (BC_1)(BC_5) &= R(BC_4 + BC_6), \\
(BC_1)(BC_6) &= R(BC_5 + BC_7), & (BC_1)(BC_7) &= R(BC_6 + BC_8), \\
(BC_2)(BC_3) &= R(BC_1 + BC_5), & (BC_2)(BC_5) &= R(BC_3 + BC_7), \\
(BC_2)(BC_6) &= R(BC_4 + BC_8), & (BC_2)(BC_7) &= R(BC_5 - BC_8), \\
(BC_4)(BC_3) &= R(BC_1 + BC_7), & (BC_4)(BC_5) &= R(BC_1 - BC_8), \\
(BC_4)(BC_6) &= R(BC_2 - BC_7), & (BC_4)(BC_7) &= R(BC_3 - BC_6), \\
(BC_8)(BC_3) &= R(BC_5 - BC_6), & (BC_8)(BC_5) &= R(BC_3 - BC_4), \\
(BC_8)(BC_6) &= R(BC_2 - BC_3), & (BC_8)(BC_7) &= R(BC_1 - BC_2).
\end{aligned}$$

Substituting we obtain

$$\begin{aligned}(M - N)(P - Q) \\ &= 4R(BC_1 - BC_2 + BC_3 - BC_4 + BC_5 - BC_6 + BC_7 - BC_8) = 4R^2.\end{aligned}$$

3.8.10 PROBLEM. *To construct a regular 17-sided polygon.*

From the above we have

$$\begin{aligned}
(M - N)(P - Q) &= 4R^2, & (M - N) - (P - Q) &= R, \\
MN &= R^2, & PQ &= R^2, \\
(BC_2)(BC_8) &= RN, & BC_2 + BC_8 &= P.
\end{aligned}$$

Knowing the product and difference of $M - N$ and $P - Q$, we can construct these two quantities. Then, knowing the product and difference of M and N, we can construct N. Similarly, knowing the product and difference of P and Q, we can construct P. Finally, knowing the product and sum of BC_2 and BC_8, we can construct these quantities, and thence the regular 17-sided polygon inscribed in a circle of radius R.

The following sequence of figures shows the actual step-by-step construction.

(1) Let $ab = R$.

Figure 3.8c

(2) Take $cd = R$, $ce = 2R$. Then $eg = M - N$ and $ef = P - Q$.

Figure 3.8d

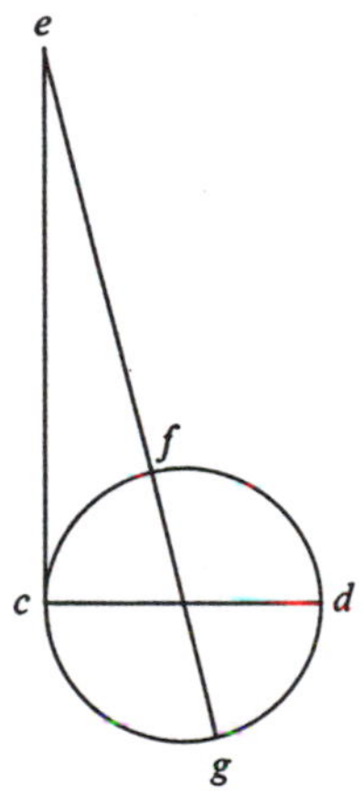

(3) Take $hi = M - N$, $hj = R$. Then $jl = M$, $jk = N$.

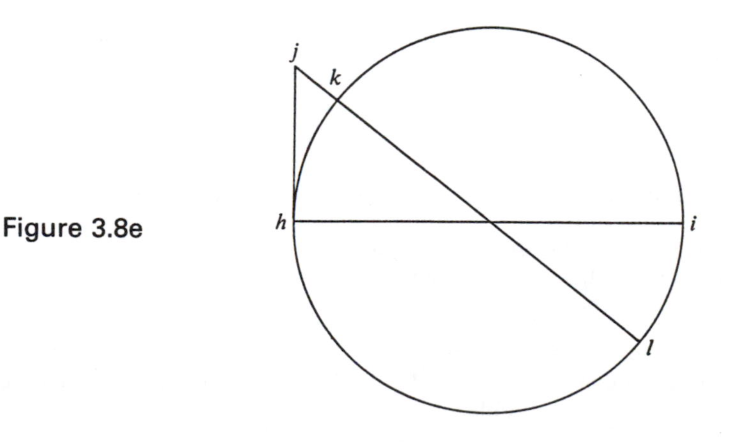

Figure 3.8e

(4) Take $mn = P - Q$, $mo = R$. Then $oq = P$, $op = Q$.

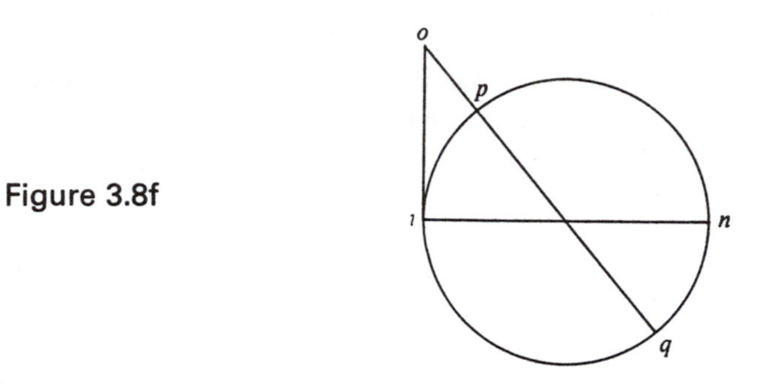

Figure 3.8f

(5) Find tu, the mean proportional between $rt = R$ and $ts = N$.

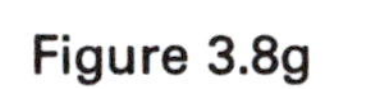
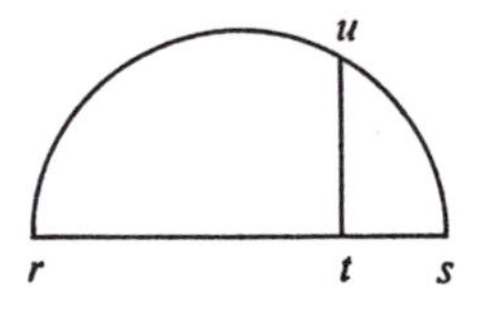

Figure 3.8g

(6) Take $vw = P$, $vx = tu$. Draw xy parallel to vw cutting semicircle on vw at y. Drop yz perpendicular to vw. Then $vz = BC_8$, $zw = BC_2$.

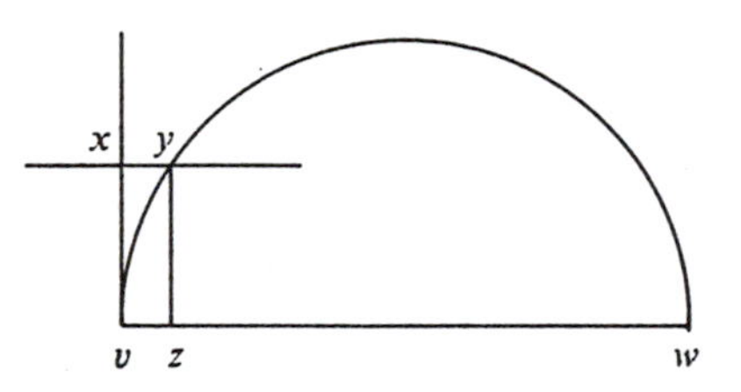

Figure 3.8h

PROBLEMS

1. In 1732, Euler showed that $f(5)$ has the factor 641. Verify this.
2. Suppose $n = rs$, where n, r, s are positive integers. Show that if a regular n-gon is constructible with Euclidean tools, then so also are a regular r-gon and a regular s-gon.
3. Suppose r and s are relatively prime positive integers and that a regular r-gon and a regular s-gon are constructible with Euclidean tools. Show that a regular rs-gon is also so constructible.

4. It can be shown that the only regular polygons that can be constructed with Euclidean tools are those the number n of whose sides can be expressed in the form

$$2^{\alpha} f(\alpha_1) f(\alpha_2) \cdots f(\alpha_k),$$

where $\alpha, \alpha_1, \alpha_2, \ldots, \alpha_k$ are distinct integers and each $f(\alpha_i)$ is a prime. On the basis of this theorem, list those regular n-gons, $n < 100$, that can be constructed with Euclidean tools.

5. Complete the proof of Theorem 3.8.5.

6. Complete the proof of Theorem 3.8.7.

7. Verify the details in the proof of Theorem 3.8.8.

8. Verify the details in the proof of Theorem 3.8.9.

9. Draw a circle and actually, with Euclidean tools, divide its circumference into 17 equal parts.

10. Show that

$$BC_8 = R\left[-1 + \sqrt{17} + \sqrt{34 - 2\sqrt{17}}\right]/8$$
$$+R\left[\sqrt{68 + 12\sqrt{17} - 16\sqrt{34 + 2\sqrt{17}} - 2(1 - \sqrt{17})\sqrt{34 - 2\sqrt{17}}}\right]/8.$$

11. Show that Theorem 3.8.2 can be generalized to hold for any regular polygon of an odd number of sides.

12. (a) Carry out a treatment of the construction of a regular pentagon analogous to the treatment in the text of the construction of a regular 17-sided polygon.
(b) Show that for the regular pentagon, $BC_2 = R(\sqrt{5} - 1)/2$.

13. Try to verify the following construction of a regular 17-gon (H. W. Richmond, "To construct a regular polygon of seventeen sides," *Mathematische Annalen*, vol. 67 (1909), p. 459).

    Let OA and OB be two perpendicular radii of a given circle with center O. Find C on OB such that $OC = OB/4$. Now find D on OA such that angle OCD = (angle OCA)/4. Next find E on AO produced such that angle DCE = 45°. Draw the circle on AE as diameter, cutting OB in F, and then draw the circle $D(F)$, cutting OA and AO produced in G_4 and G_6. Erect perpendiculars to OA at G_4 and G_6, cutting the given circle in P_4 and P_6. These last points are the fourth and sixth vertices of the regular 17-gon whose first vertex is A.

14. Show that the smaller acute angle in a right triangle with legs 3 and 16 is very closely half the central angle subtended by one side of a regular 17-gon. Using this fact, give an approximate Euclidean construction of a regular 17-gon.

3.9 THE IMPOSSIBILITY OF SOLVING THE THREE FAMOUS PROBLEMS OF ANTIQUITY WITH EUCLIDEAN TOOLS (OPTIONAL)

It was not until the nineteenth century that the three famous problems of antiquity (the duplication of a cube, the trisection of a general angle, and the quadrature of a circle) were finally shown to be impossible with Euclidean tools. The needed criterion for determining whether a given construction is or is not within the power of the tools turned out to be essentially algebraic in nature. In particular, the following two theorems (which we shall accept without proof) were established.*

3.9.1 THEOREM. *From a given unit length it is impossible to construct, with Euclidean tools, a line segment the magnitude of whose length is a root of a cubic equation with rational coefficients but with no rational root.*

3.9.2 THEOREM. *The magnitude of any length constructible with Euclidean tools from a given unit length is an algebraic number.*†

With the aid of the above two theorems we will dispose of the three famous problems of antiquity.

3.9.3 THEOREM. *The problem of duplicating a cube—that is, of constructing the edge of a cube having twice the volume of a given cube—is impossible with Euclidean tools.*

Let us take as our unit of length the edge of the given cube and let x denote the edge of the sought cube. Then we have $x^3 = 2$, or $x^3 - 2 = 0$. It follows that if a cube can be duplicated with Euclidean tools, we can, with these tools, construct from a unit segment a segment of length x. But this is impossible, by Theorem 3.9.1, since the cubic equation $x^3 - 2 = 0$ has rational coefficients but no rational root.‡

3.9.4 THEOREM. *The problem of trisecting a general angle is impossible with Euclidean tools.*

* See, for example, Howard Eves, *The Foundations and Fundamental Concepts of Mathematics*, 3d ed. (Boston: PWS-KENT, 1990), pp. 276–284.

† A number is said to be *algebraic* if it is a root of a polynomial equation having rational coefficients.

‡ It is known from algebra that if a polynomial equation

$$a_0x^n + a_1x^{n-1} + \cdots + a_n = 0,$$

with integral coefficients $a_0, a_1, \ldots, a_n$, has a reduced rational root a/b, then a is a factor of a_n and b is a factor of a_0. Thus any rational roots of $x^3 - 2 = 0$ are among $1, -1, 2, -2$. Since direct testing shows that none of these numbers satisfies the equation, the equation has no rational roots.

We may show that the *general* angle cannot be trisected with Euclidean tools by showing that some *particular* angle cannnot be so trisected. Now, from trigonometry, we have the identity

$$\cos \theta = 4 \cos^3(\theta/3) - 3 \cos(\theta/3).$$

If we take $\theta = 60°$ and set $x = \cos(\theta/3)$, this becomes

$$8x^3 - 6x - 1 = 0.$$

Figure 3.9a

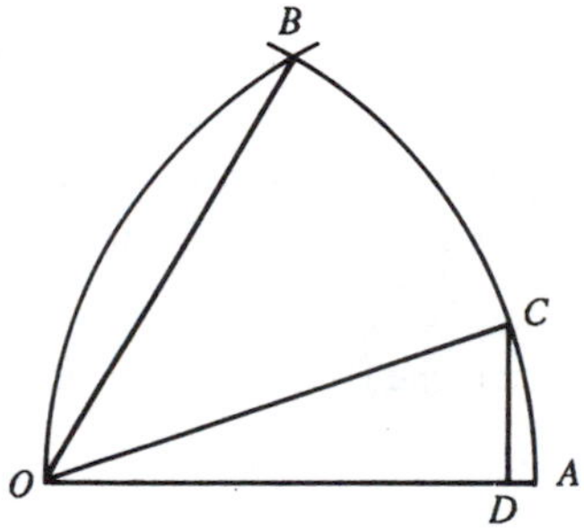

Let OA (see Figure 3.9a) be a given unit segment. Describe the circle $O(A)$ and then the circle $A(O)$ to intersect in B. Then angle $BOA = 60°$. Let trisector OC, which makes angle $COA = 20°$, cut the circle OA in C, and let D be the foot of the perpendicular from C on OA. Then $OD = \cos 20° = x$. It follows that if a 60° angle can be trisected with Euclidean tools—in other words, if OC can be drawn with these tools—we can construct from a unit segment OA another segment of length x. But this is impossible, by Theorem 3.9.1, since the above cubic equation has rational coefficients but no rational root.

It should be noted that we have not proved that *no* angle can be trisected with Euclidean tools, but only that *not all* angles can be so trisected. The truth of the matter is that the 90° angle and an infinite number of other angles can be trisected by the use of Euclidean tools.

3.9.5 THEOREM. *The problem of squaring a circle—that is, of constructing a square whose area is equal to that of a given circle—is impossible with Euclidean tools.*

Let us take as our unit of length the radius of the given circle and let x denote the side of the sought square. Then we have $x = \sqrt{\pi}$. It follows that if we can square a circle with Euclidean tools, we can, with these tools, construct from a unit segment a segment of length $\sqrt{\pi}$. But this is impossible, by Theorem 3.9.2, since π, and hence also $\sqrt{\pi}$, was shown by C. L. F. Lindemann in 1882 to be nonalgebraic.

PROBLEMS

1. Show that it is impossible with Euclidean tools to construct a regular 9-sided polygon.

2. Show that it is impossible with Euclidean tools to construct an angle of $1°$.

3. Show that it is impossible with Euclidean tools to construct a regular 7-sided polygon.

4. Establish the trigonometric identity used in the proof of Theorem 3.9.4.

5. Show that it is impossible with Euclidean tools to trisect an angle whose cosine is $\frac{2}{3}$.

6. Given a segment s, show that it is impossible with Euclidean tools to construct segments m and n such that $s:m = m:n = n:2s$.

7. Show that it is impossible with Euclidean tools to construct a line segment whose length equals the circumference of a given circle.

8. Let $OADB$ be an arbitrary given rectangle. Show that it is impossible with Euclidean tools to draw a circle, concentric with the rectangle, cutting OA and OB produced in A' and B' such that A', D, B' are collinear.

9. Let AB be a given segment. Draw angle $ABM = 90°$ and angle $ABN = 120°$, both angles lying on the same side of AB. Show that it is impossible with Euclidean tools to draw a line ACD cutting BM in C and BN in D such that $CD = AB$.

10. Let AOB be a central angle in a given circle. Show that it is impossible with Euclidean tools to draw a line BCD through B cutting the circle again in C, AO produced in D, and such that $CD = OA$.

11. An angle AOB and a point P within the angle are given. The line through P cutting OA and OB in C and D so that $CE = PD$, where E is the foot of the perpendicular from O on CD, is known as *Philon's line for angle AOB and the point P*. It can be shown that Philon's line is the minimum chord CD that can be drawn through P. Show that in general it is impossible to construct with Euclidean tools Philon's line for a given angle and a given point.

BIBLIOGRAPHY

ALTSHILLER-COURT, NATHAN, *College Geometry, an Introduction to the Modern Geometry of the Triangle and the Circle*. New York: Barnes and Noble, 1952.

COOLIDGE, J. L., *A Treatise on the Circle and the Sphere*. New York: Oxford University Press, 1916.

______, *A History of Geometrical Methods*. New York: Oxford University Press, 1940.

DAUS, P. M., *College Geometry*. Englewood Cliffs, N.J.: Prentice-Hall, 1941.

DAVIS, D. R., *Modern College Geometry*. Reading, Mass.: Addison-Wesley, 1949.

DICKSON, L. E., *New First Course in the Theory of Equations*. New York: John Wiley and Sons, 1939.

EVES, HOWARD, *An Introduction to the History of Mathematics*, 6th ed. Philadelphia: Saunders College Publishing, 1990.

______, *The Foundations and Fundamental Concepts of Mathematics*, 3d ed. Boston: PWS-KENT, 1990.

HUDSON, H. P., *Ruler and Compasses*. New York: Longmans, Green and Company, 1916. Reprinted in *Squaring the Circle, and Other Monographs*, Chelsea Publishing Company, 1953.

JARDIN, DOV., *Constructions with the Bi-Ruler and with the Double-Ruler*. Jerusalem, Israel: privately printed, 1964.

JOHNSON, D. A., *Paper Folding for the Mathematics Class*. Washington, D.C.: National Council of Teachers of Mathematics, 1957.

KAZARINOFF, N. D., *Ruler and the Round*. Boston: Prindle, Weber & Schmidt, 1970.

KEMPE, A. B., *How to Draw a Straight Line*; *a Lecture on Linkages*. New York: The Macmillan Company, 1887. Reprinted in *Squaring the Circle, and Other Monographs*, Chelsea Publishing Company, 1953.

KLEIN, FELIX, *Famous Problems of Elementary Geometry*, tr. by W. W. Beman and D. E. Smith. New York: G. E. Stechert and Company, 1930. Reprinted in *Famous Problems, and Other Monographs*, Chelsea Publishing Company, 1955. Original German volume published in 1895.

KOSTOVSKII, A. N., *Geometrical Constructions Using Compasses Only*, tr. by Halina Moss. New York: Blaisdell Publishing Company, 1961. Original Russian volume published in 1959.

MESCHOWSKI, HERBERT, *Unsolved and Unsolvable Problems in Geometry*, tr. by J. A. C. Burlak. Edinburgh: Oliver and Boyd, 1968.

MILLER, L. H., *College Geometry*. New York: Appleton-Century-Crofts, 1957.

MOISE, E. E., *Elementary Geometry from an Advanced Standpoint*, 3d ed. Reading, Mass.: Addison-Wesley, 1990.

PETERSEN, JULIUS, *Methods and Theories for the Solution of Problems of Geometrical Construction Applied to 410 Problems*, tr. by Sophus Haagensen. New York: G. E. Stechert and Company, 1927. Reprinted in *String Figures, and Other Monographs*, Chelsea Publishing Company, 1960. Original Danish volume published in 1879.

RANSOM, W. R., *Can and Can't in Geometry*. Portland, Me.: J. Weston Walsh, 1960.

ROW, SUNDARA, *Geometric Exercises in Paper Folding*, tr. by W. W. Beman and D. E. Smith. Chicago: The Open Court Publishing Company, 1901. Original volume published in Madras in 1893.

SHIVELY, L. S., *An Introduction to Modern Geometry*. New York: John Wiley and Sons, 1939.

SMOGORZHEVSKII, A. S., *The Ruler in Geometrical Constructions*, tr. by Halina Moss. New York: Blaisdell Publishing Company, 1961. Original Russian volume published in 1957.

STEINER, JACOB, *Geometrical Constructions with a Ruler Given a Fixed Circle with Its Center*, tr. by M. E. Stark. New York: Scripta Mathematica, 1950. First German edition published in 1833.

YATES, R. C., *Geometrical Tools, a Mathematical Sketch and Model Book*. Saint Louis: Educational Publishers, 1949.

______, *The Trisection Problem*. Ann Arbor, Mich.: Edwards Brothers, 1947.

YOUNG, J. W. A., ed., *Monographs on Topics of Modern Mathematics Relevant to the Elementary Field*. New York: Longmans, Green and Company, 1911. Reprinted by Dover Publications, 1955.

Appendix: Perspectivity Transformation (Optional)

Many theorems of modern elementary geometry can be neatly established by transformations other than the *elementary transformations* treated in Chapter 2. Particularly useful among these other transformations is the so-called *perspectivity transformation*. This appendix is devoted to a brief consideration of this transformation.

Figure Aa

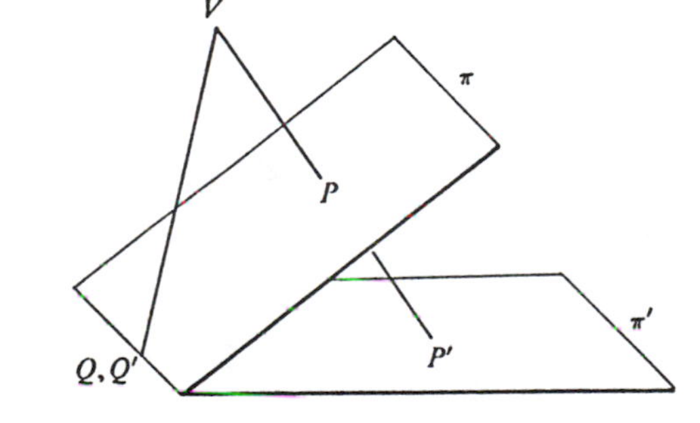

Let π and π' (see Figure Aa) be two given fixed nonideal planes of *extended* space, and let V be a given fixed point not lying on either π or π'. Since the space is extended, π and π' are extended planes, and point V may be either an ordinary or an ideal point. Let P be any point, ordinary or ideal, of plane π. Then line VP will intersect plane π' in a unique ordinary or ideal point P' of π'. In this way the extended plane π is mapped onto the extended plane π'. Indeed, since distinct points of π have distinct images in π', the mapping is actually a transformation of the set of all points of the extended plane π onto the set of all points of the extended plane π'. The points on the line of intersection of planes π and π' are invariant points of the transformation.

A.1 Definitions. A transformation such as described above is called a *perspectivity transformation*, and the point V is called the *center* of the perspectivity. If V is an ordinary point of space, the perspectivity is called a *central perspectivity*; if V is an ideal point of space, the perspectivity is called a *parallel perspectivity*. The line of intersection of π and π' is called the *axis of perspectivity*. The line in $\pi(\pi')$ that maps into the line at infinity in $\pi'(\pi)$ is called the *vanishing line* of $\pi(\pi')$. The point in which a line of $\pi(\pi')$ meets the vanishing line of $\pi(\pi')$ is called the *vanishing point* of the line.

It is clear that the two planes π and π' of a perspectivity must be taken as *extended* planes, since otherwise the correspondence between the points of the two planes might not be one-to-one. It is for this reason that an extended plane is often called a *projective plane*.

It is easily seen that a perspectivity carries a straight line into a straight line. Very useful is the following special situation.

A.2 THEOREM. *If π is a given plane, l a given line in π, and V a given point not on π, then there exists a plane π' such that the perspectivity of center V carries line l of π into the line at infinity of π'.*

Figure Ab

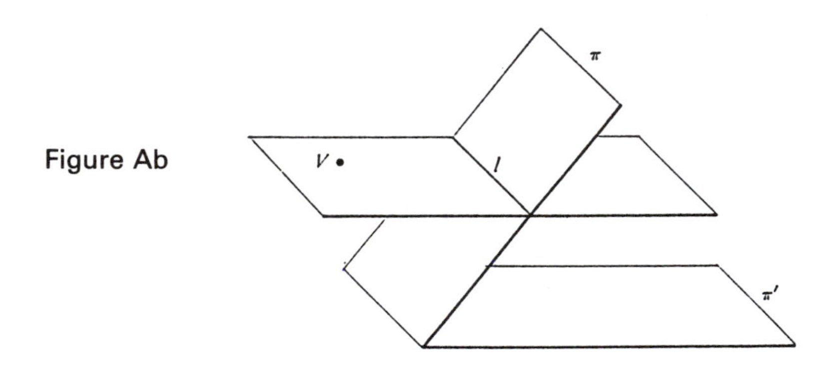

Choose for π' (see Figure Ab) any plane parallel to (but not coincident with) the plane determined by V and l. Then it is clear that the line joining V to any point P on l will be parallel to plane π' (that is, will meet π' at infinity).

A.3 DEFINITION. The operation of selecting a suitable center of perspectivity V and a plane π' so that a given line l of a given plane π shall be mapped into the line at infinity of π' is called *projecting the given line to infinity*.

The operation of projecting a given line to infinity can often greatly simplify the proof of a theorem. We give some examples; the reader should note the application of the *transform-solve-invert* procedure. We first establish one of the harmonic properties of a complete quadrangle.

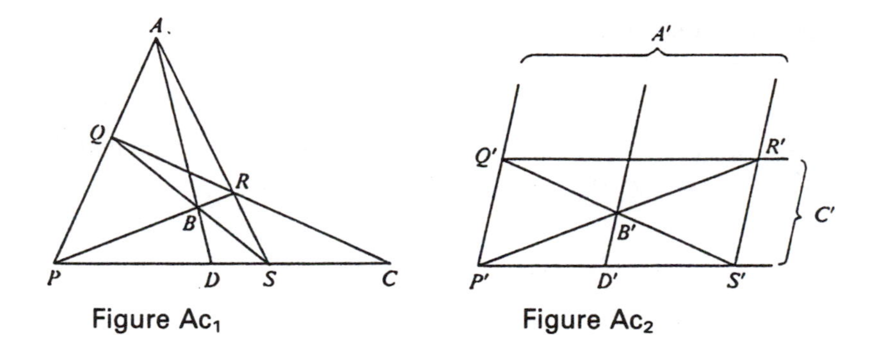

Figure Ac_1 Figure Ac_2

A.4 THEOREM. *Let PQRS (see Figure Ac_1) be a complete quadrangle, and let PQ and SR intersect in A, PR and SQ intersect in B, PS and QR intersect in C, AB and PS intersect in D. Then $(PS, DC) = -1$.*

Project line AC to infinity. Then $P'Q'R'S'$ (see Figure Ac_2) is a parallelogram and D' is the midpoint of $P'S'$. Since C' is at infinity, we have $(P'S', D'C') = -1$. The theorem now follows.

A.5 THEOREM. *If the three lines UP_1P_2, UQ_1Q_2, UR_1R_2 (see Figure*

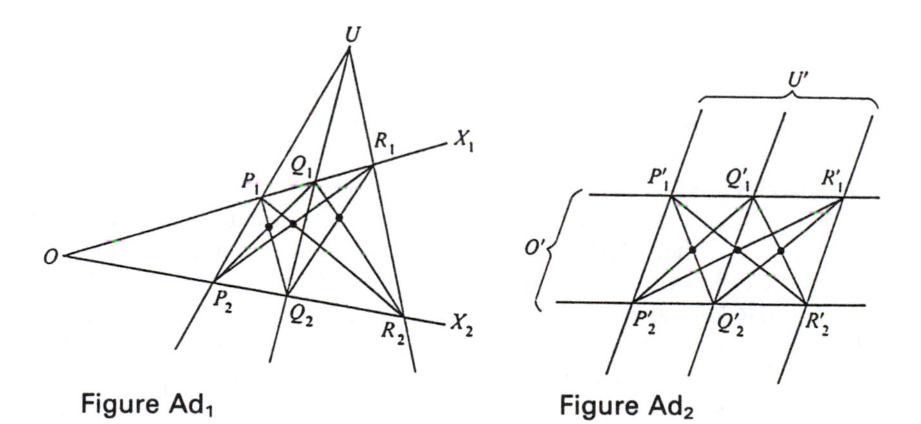

Figure Ad_1 Figure Ad_2

Ad_1) intersect the two lines OX_1 and OX_2 in P_1, Q_1, R_1 and P_2, Q_2, R_2 respectively, then the points of intersection of Q_1R_2 and Q_2R_1, R_1P_2 and R_2P_1, P_1Q_2 and P_2Q_1 are collinear on a line that passes through point O.

Project line OU to infinity. The projected figure appears as in Figure Ad_2, where $P_1'P_2'$, $Q_1'Q_2'$, $R_1'R_2'$ are all parallel, and $P_1'Q_1'R_1'$ and $P_2'Q_2'R_2'$ are parallel. It is clear that the points of intersection of $Q_1'R_2'$ and $Q_2'R_1'$, $R_1'P_2'$ and $R_2'P_1'$, $P_1'Q_2'$ and $P_2'Q_1'$ are collinear on a line parallel to lines $P_1'Q_1'R_1'$ and $P_2'Q_2'R_2'$ (they lie on the line midway between lines $P_1'Q_1'R_1'$

and $P_2'Q_2'R_2'$). It follows that the corresponding points in the original figure are collinear on a line passing through point O.

A.6 DESARGUES' TWO-TRIANGLE THEOREM. *Copolar triangles in a plane are coaxial, and conversely.*

Figure Ae

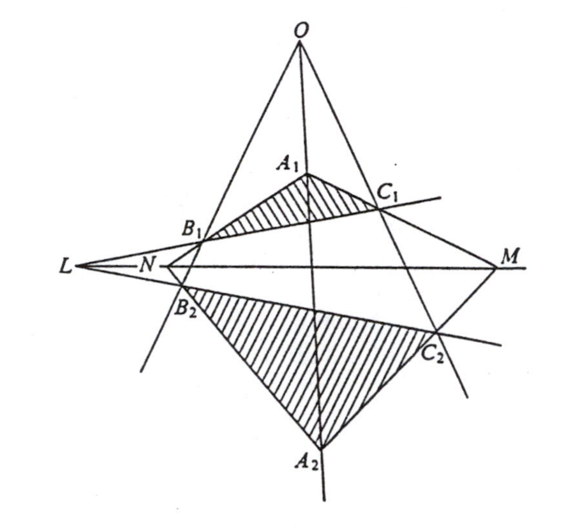

Let the two triangles (see Figure Ae) be $A_1B_1C_1$ and $A_2B_2C_2$, and let B_1C_1 and B_2C_2 intersect in L, C_1A_1 and C_2A_2 intersect in M, A_1B_1 and A_2B_2 intersect in N.

Suppose A_1A_2, B_1B_2, C_1C_2 are concurrent in a point O. Project line MN to infinity. Then $A_1'B_1'$ and $A_2'B_2'$ are parallel, and $A_1'C_1'$ and $A_2'C_2'$ are parallel. It follows that $O'B_1'/O'B_2' = O'A_1'/O'A_2' = O'C_1'/O'C_2'$, whence $B_1'C_1'$ and $B_2'C_2'$ are also parallel. That is, the intersections of corresponding sides of triangles $A_1'B_1'C_1'$ and $A_2'B_2'C_2'$ are collinear (on the line at infinity). It follows that the intersections of corresponding sides of triangles $A_1B_1C_1$ and $A_2B_2C_2$ are collinear. That is, copolar triangles in a plane are coaxial.

Now suppose L, M, N are collinear. Project line LMN to infinity. Then the two triangles $A_1'B_1'C_1'$ and $A_2'B_2'C_2'$ have their corresponding sides parallel, and hence are homothetic to one another, whence $A_1'A_2'$, $B_1'B_2'$, $C_1'C_2'$ are concurrent in a point O'. It follows that A_1A_2, B_1B_2, C_1C_2 are concurrent in a point O. That is, coaxial triangles in a plane are copolar.

A.7 THEOREM. *There is a perspectivity that carries a given triangle ABC and a given point G in its plane (but not on a side line of the triangle) into a triangle $A'B'C'$ and its centroid G'.*

Figure Af

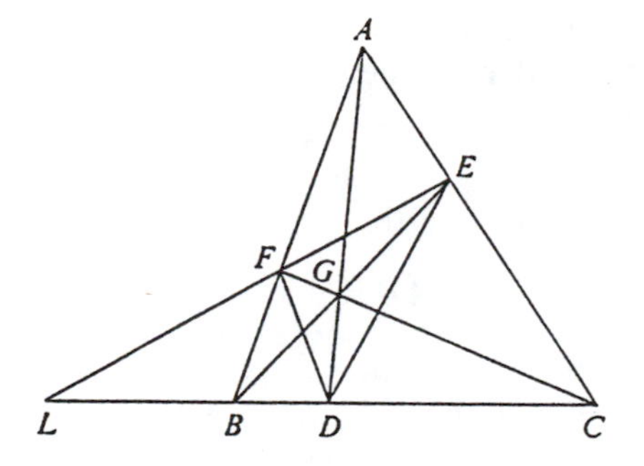

Let AG, BG, CG (see Figure Af) intersect the opposite sides BC, CA, AB in points D, E, F. Then triangles DEF and ABC are copolar, and hence also coaxial. That is, the points L, M, N of intersection of EF and BC, FD and CA, DE and AB are collinear. By Theorem A.4, $(BC, DL) = -1$. Project line LMN to infinity. Then D' is the midpoint of $B'C'$. Similarly E' and F' are the midpoints of $C'A'$ and $A'B'$ respectively. It follows that G' is the centroid of triangle $A'B'C'$.

If AD, BE, CF are concurrent cevian lines for a triangle ABC, Ceva's Theorem states that

$$\frac{(\overline{BD})(\overline{CE})(\overline{AF})}{(\overline{DC})(\overline{EA})(\overline{FB})} = +1.$$

The left member of this equation is a ratio of a product of some directed segments to a product of some other directed segments, and this ratio of products of directed segments has the two following interesting properties:

(1) If we replace $\overline{BD}$ by $B \times D$, and similarly treat all other segments appearing, and then regard the resulting expression as an algebraic one in the letters B, D, etc., these letters can all be cancelled out.

(2) If we replace BD by a letter, say a, representing the line on which the segment is found, and similarly treat all other segments appearing, and then regard the resulting expression as an algebraic one in the letters a, etc., these letters can all be cancelled out.

A.8 Definition. A ratio of a product of directed segments to another product of directed segments, where all the segments lie in one plane, is called an *h-expression* if it has the properties (1) and (2) described above.

A.9 Theorem. *The value of an h-expression is invariant under any perspectivity.*

It is sufficient to show that a given h-expression has a value that is invariant under an arbitrary perspectivity. Let V be the center of a perspectivity, and let $\overline{AB}$ be any one of the segments appearing in the h-expression. Let p denote the perpendicular distance from V to AB. Then

$$p\overline{AB} = (VA)(VB) \sin \overline{AVB},$$

since each side is twice the area of $\triangle\overline{VAB}$. It follows that

$$\overline{AB} = [(VA)(VB) \sin \overline{AVB}]/p.$$

Replace each $\overline{AB}$ in the h-expression by $[(VA)(VB) \sin \overline{ABV}]/p$. Since property (1) above holds, VA, VB, etc., cancel out; since property (2) above holds, p, etc., cancel out. We are left with an expression containing only the sines $\sin \overline{AVB}$, etc. Now, since $\sphericalangle\overline{AVB} = \sphericalangle\overline{A'VB'}$, etc., the same relation that holds among the sines $\sin \overline{AVB}$, etc., holds among the sines $\sin \overline{A'VB'}$, etc. Introducing factors VA', VB', p', etc., by reversing the earlier cancellations, leads to the same relation among segments $\overline{A'B'}$, etc., as was given among segments $\overline{AB}$, etc.

The reader will note that the procedure employed in the above proof is essentially the way we proved, in Section 1.5, that the cross ratio of four collinear points is invariant under a perspectivity.

A.10 CEVA'S THEOREM. *If AD, BE, CF are concurrent cevian lines for a triangle ABC, then*

$$\frac{(\overline{BD})(\overline{CE})(\overline{AF})}{(\overline{DC})(\overline{EA})(\overline{FB})} = +1. \qquad (1)$$

The expression on the left of (1) is an h-expression and is therefore (by Theorem A.9) invariant in value under projection. By Theorem A.7, there is a perspectivity that carries triangle ABC and the point G of concurrence of the three cevian lines into a triangle $A'B'C'$ and its centroid G'. Then $\overline{B'D'}/\overline{D'C'} = \overline{C'E'}/\overline{E'A'} = \overline{A'F'}/\overline{F'B'} = 1$, and clearly

$$\frac{(\overline{B'D'})(\overline{C'E'})(\overline{A'F'})}{(\overline{D'C'})(\overline{E'A'})(\overline{F'B'})} = +1.$$

The theorem now follows.

A.11 MENELAUS' THEOREM. *If D, E, F are collinear menelaus points on the sides BC, CA, AB of a triangle ABC, then*

$$\frac{(\overline{BD})(\overline{CE})(\overline{AF})}{(\overline{DC})(\overline{EA})(\overline{FB})} = -1. \qquad (1)$$

The expression on the left of (1) is an h-expression and is therefore (by Theorem A.9) invariant in value under projection. Project line DEF to infinity. Then $\overline{B'D'}/\overline{D'C'} = \overline{C'E'}/\overline{E'A'} = \overline{A'F'}/\overline{F'B'} = -1$, and clearly

$$\frac{(\overline{B'D'})(\overline{C'E'})(\overline{A'F'})}{(\overline{D'C'})(\overline{E'A'})(\overline{F'B'})} = -1.$$

The theorem now follows.

We conclude the section by giving another projective proof of Desargues' Two-Triangle Theorem and a proof of a theorem due to Pappus. The latter theorem is a generalization of Theorem A.5.

A.12 DESARGUES' TWO-TRIANGLE THEOREM. *Copolar triangles (in space or in a plane) are coaxial, and conversely.*

Figure Ag

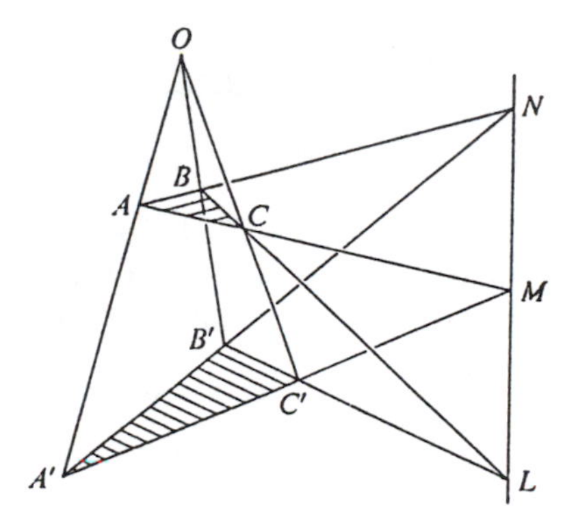

(a) Let the two triangles be ABC and $A'B'C'$, and suppose they lie in different planes π and π' respectively (see Figure Ag). Also suppose AA', BB', CC' are concurrent in a point O. Then BC and $B'C'$ are coplanar and thus intersect in a point L. Similarly, CA and $C'A'$ are coplanar and thus intersect in a point M, and AB and $A'B'$ are coplanar and thus intersect in a point N. The points L, M, N then lie in both planes π and π', and hence on the line of intersection of these two planes. That is, copolar triangles in different planes are coaxial. The converse follows by reversing the above argument.

(b) Now suppose the two planes π and π' coincide, and (see Figure Ah) that the points L, M, N of intersection of BC and $B'C'$, CA and $C'A'$, AB and $A'B'$ are collinear on a line l in π. Let π_1 be a plane distinct from π and passing through l, let P be a point not on π or π_1, and let PA', PB', PC' cut π_1 in A_1, B_1, C_1, respectively. Then B_1, C_1, B', C' are coplanar, as also are C_1, A_1, C', A' and A_1, B_1, A', B', and we see that BC, $B'C'$, B_1C_1 meet in L; CA, $C'A'$, C_1A_1 meet in M; AB, $A'B'$, A_1B_1 meet in N. Thus the two triangles $A_1B_1C_1$ and ABC are coaxial and therefore, by part (a), are copolar. That is, AA_1, BB_1, CC_1 are concurrent in a point Q. Let QP cut π in point O. Then A, A', O all lie in the plane

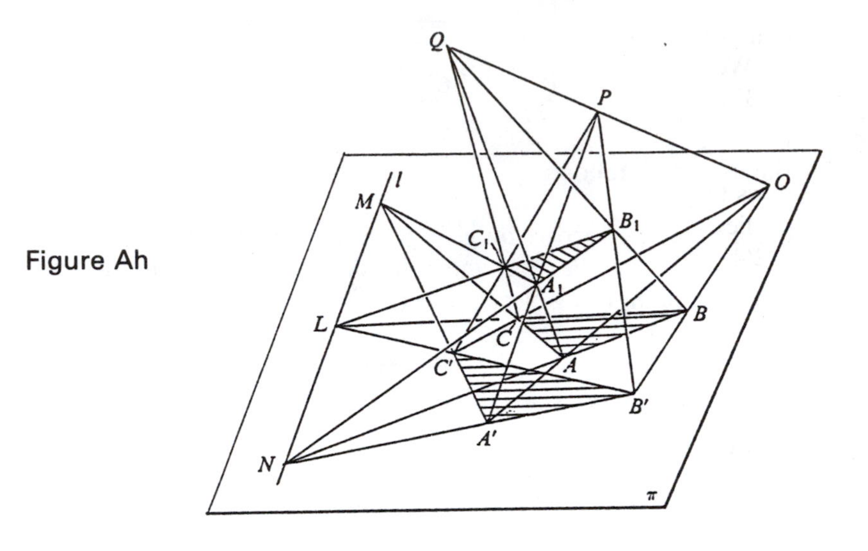

Figure Ah

determined by PA_1A' and QA_1A. It follows that AA' passes through O. Similarly, BB' and CC' pass through O, and triangles ABC and $A'B'C'$ are copolar. The converse is obtained by applying the above to triangles $AA'M$ and $BB'L$.

A.13 PAPPUS' THEOREM. *If the vertices 1, 2, 3, 4, 5, 6 of a hexagon 123456 lie alternately on a pair of lines, then the three intersections P, Q, R of the opposite sides 23 and 56, 45 and 12, 61 and 34 of the hexagon are collinear (see Figure Ai_1).*

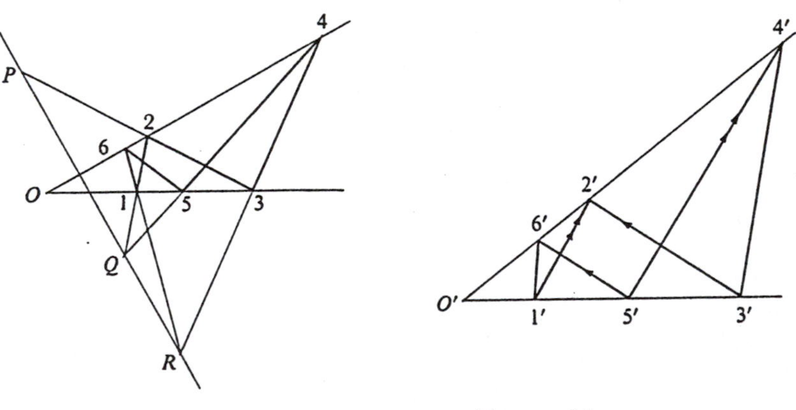

Figure Ai_1

Figure Ai_2

Project line PQ to infinity, and suppose the intersection O of lines 135 and 246 does not lie on PQ. Then 1′, 3′, 5′ are collinear, 2′, 4′, 6′ are collinear, O' is a finite point, $2'3'$ is parallel to $5'6'$, and $4'5'$ is parallel to $1'2'$. We must prove that $6'1'$ is parallel to $3'4'$. Now (see Figure Ai_2), since $2'3'$ is parallel to $5'6'$ and $4'5'$ is parallel to $1'2'$,

$$\overline{O'6'}/\overline{O'2'} = \overline{O'5'}/\overline{O'3'} \text{ and } \overline{O'1'}/\overline{O'5'} = \overline{O'2'}/\overline{O'4'}.$$

It follows that $\overline{O'6'}/\overline{O'1'} = \overline{O'4'}/\overline{O'3'}$ and $\overline{6'1'}$ is parallel to $\overline{3'4'}$.

If O lies on PQ, then lines $1'3'5'$ and $2'4'6'$ are parallel. But then $\overline{1'5'} = \overline{2'4'}$ and $\overline{5'3'} = \overline{6'2'}$. It follows that $\overline{1'3'} = \overline{6'4'}$ and $6'1'$ is parallel to $3'4'$.

PROBLEMS

1. (a) Prove that a perspectivity carries a straight line into a straight line.
   (b) Prove that under a perspectivity a straight line and its image intersect each other on the axis of perspectivity.
2. (a) Prove that under a perspectivity the angle between the images m' and n' of any two lines m and n is equal to the angle that the vanishing points of m and n subtend at the center V of perspectivity.
   (b) Prove that under a perspectivity all angles whose sides have the same vanishing points map into equal angles.
3. (a) Suppose a plane π contains a line l and two angles ABC and DEF, where A, C, D, F are on l in the order A, D, C, F. Show that there exists a perspectivity that projects l to infinity and angles ABC and DEF into angles of given sizes α and β respectively.
   (b) Must the segments AC and DF in part (a) separate each other?
4. Show that there exists a perspectivity that projects a given quadrilateral $ABCD$ into a square.
5. Prove that the cross ratio of a pencil of four distinct coplanar lines is preserved by projection.
6. Show that in a perspectivity carrying a plane π into a nonparallel plane π' there are in each plane exactly two points such that every angle at either of them projects into an equal angle. (These points are called the *isocenters* of the perspectivity.)
7. Show that in a perspectivity carrying a plane π into a nonparallel plane π' there is in each plane, besides the axis of perspectivity, a line whose segments are projected into equal segments. (These lines are called the *isolines* of the perspectivity.)

Suggestions for Solutions of Selected Problems

Section 1.1

1. Use Theorem 1.1.3 along with mathematical induction.

2. Start with $\overline{AM} = \overline{MB}$ and then insert an origin at P.

3. Start with $\overline{AB} = \overline{OB} - \overline{OA}$ and then square both sides.

4. Insert an origin at P.

5. Set $\overline{AA'} = \overline{OA'} - \overline{OA} = (\overline{O'A'} - \overline{O'O}) - \overline{OA}$, etc.

6. Insert an origin O and let M and N denote the midpoints of CR and PQ. Then $4\overline{OM} = 2\overline{OR} + 2\overline{OC} = \overline{OA} + \overline{OB} + 2\overline{OC} = \overline{OB} + \overline{OC} + 2\overline{OQ} = 2\overline{OP} + 2\overline{OQ} = 4\overline{ON}$. Or, M and N clearly coincide if A, B, C are not collinear; now let C approach collinearity with A and B.

9. Insert an origin at O.

10. Use Theorem 1.1.9.

11. Use Theorem 1.1.9.

13. First consider the case where P is on the line and take P as origin; then let P' be the foot of the perpendicular dropped from P to the line.

14. Use Stewart's Theorem.

15. Use Stewart's Theorem.

16. By Problem 15, Section 1.1, we know that $t_a^2 = bc[1 - a^2/(b+c)^2]$, $t_b^2 = ca[1 - b^2/(c+a)^2]$. Show that $t_a^2 - t_b^2 = (b-a)f(a,b,c)$ where $f(a,b,c)$ contains only positive terms.

17. Use Stewart's Theorem.

18. This problem is a generalization of Stewart's Theorem in that the point P is replaced by a circle.

19. (a) Let a line cut OA, OB, OC, OD in A', B', C', D'. Apply Euler's Theorem to A', B', C', D'. Multiply through by p^2, where p is the perpendicular from

O onto line $A'B'C'D'$. Replace $p\overline{A'B'}$ by $(OA')(OB')\sin \overline{A'OB'}$, etc. (b) $2\triangle\overline{AOB} = (OA)(OB)\sin\overline{AOB}$, etc. Now use Problem 19(a), Section 1.1.

20. Take O at the center of the circle or on the circumference of the circle.

21. Consider four cases: first where O is within triangle ABC, next where O is within $\sphericalangle BAC$ but on the other side of BC from A, then where O is within the vertical angle of $\sphericalangle BAC$, finally where O is on a side of triangle ABC.

23. Consider the equation where A is replaced by a variable point X on the line. Take any origin O on the line and obtain a quadratic equation in $x = OX$. Take $X = B, C, D$ in turn, obtaining three identities. Then the quadratic equation has three distinct roots, and therefore is an identity. That is, X can be any point on the line, e.g., A.

Section 1.2

3. Each follows from Theorem 1.2.5 or Convention 1.2.4. Thus Theorem (a) becomes: "Through a given point there is one and only one plane containing the ideal line of a given plane not passing through the given point," and this is an immediate consequence of Theorem 1.2.5. Theorem (b) becomes: "Two (distinct) lines which intersect a third line in its ideal point, intersect each other in their ideal points," and this follows from Convention 1.2.4.

5. Multiply the identity of Problem 23, Section 1.1 by $\overline{AD}$ and then take D as an ideal point.

6. Divide the identity of Problem 23, Section 1.1 by $\overline{AQ}$ and then take Q as an ideal point.

Section 1.3

2. $\overline{BD}/\overline{DC} = \overline{BF}/\overline{FL}$, etc.

3. If A', B', C' are the feet of the perpendiculars and if DEF cuts $A'B'C'$ in T, then $\overline{BD}/\overline{DC} = \overline{B'T}/\overline{TC'}$, etc.

5. Use Menelaus' Theorem.

6. $PD/AD = \triangle BPC/\triangle BAC$, etc.

7. $AP/AD = (AD - PD)/AD = 1 - PD/AD$, etc. Now use Problem 6, Section 1.3.

8. $\overline{AF}/\overline{FB} = \triangle\overline{CAP}/\triangle\overline{BCP}$, $\overline{AE}/\overline{EC} = \triangle\overline{PAB}/\triangle\overline{BCP}$.

9. Apply Menelaus' Theorem to triangles ABD and BCD.

10. Use mathematical induction. The converse is not true. A correct converse is that if the relation holds and if $n - 1$ of the points $A', B', C', D', \ldots$ are collinear, then all n of them are collinear.

11. Apply Menelaus' Theorem to triangles ABD and BCD.

12. Let the plane cut line DB in point P. Use Menelaus' Theorem on triangle ABD with transversal $D'A'P$ and on triangle CBD with transversal $C'B'P$.

13. Use Theorem 1.1.9.

14. In the figure for Problem 13, Section 1.3, draw a sphere with center O to cut OA', OB', OC', OD', OE', OF' in A, B, C, D, E, F.

15. $\overline{AA'}/\overline{A'B} = (OA \sin \overline{AOA'})/(OB \sin \overline{A'OB})$, etc.

Section 1.4

1. (b) Use the trigonometric form of Ceva's Theorem.
   (c) Use the trigonometric form of Ceva's Theorem.

2. Use Ceva's Theorem.

3. Use Ceva's Theorem.

4. Use Ceva's Theorem.

6. Use Menelaus' Theorem.

7. Use the trigonometric form of Ceva's Theorem.

8. Use the trigonometric form of Menelaus' Theorem.

9. Use Ceva's Theorem along with the fact that $(AF)(AF') = (AE)(AE')$, etc.

10. Let D, E, F be the points of intersection of the tangents at A, B, C with the opposite sides. Then triangles ABD and CAD are similar, and $\overline{BD}/\overline{DC} = -\overline{BD}/\overline{CD} = -(BD)^2/(\overline{BD})(\overline{CD}) = -(BD)^2/(AD)^2 = -c^2/b^2$, etc. Or obtain the result as a special case of Pascal's "mystic hexagram" theorem, by regarding ABC as a hexagon with two vertices coinciding at A, two at B, and two at C. Or use the fact that, by Problem 2, Section 1.4, the given triangle and the triangle formed by the tangents are copolar.

11. Use Desargues' two-triangle theorem.

12. Use the trigonometric form of Menelaus' Theorem, or use Problem 12, Section 1.1.

13. Use the trigonometric form of Menelaus' Theorem.

14. Let DD', $A'B$ intersect in E and $A'D'$, BD in G. Then, using Menelaus' Theorem on triangle $GA'B$ and transversal $D'DE$ we find

$$(\overline{BD}/\overline{DG})(\overline{GD'}/\overline{D'A'})(\overline{A'E}/\overline{EB}) = -1.$$

But $\overline{BD} = \overline{CA}$, $\overline{GD'} = \overline{BB'}$, $\overline{DG} = \overline{AA'}$, $\overline{D'A'} = \overline{B'C}$. Therefore

$$(\overline{CA}/\overline{AA'})(\overline{A'E}/\overline{EB})(\overline{BB'}/\overline{B'C}) = -1.$$

15. Apply Menelaus' Theorem to triangle ABD with transversal EG and to triangle CBD with transversal HF.

16. Triangles CAC' and $B'AB$ are congruent and therefore $\sin ACC'/\sin B'BA = \sin ACC'/\sin CC'A = AC'/CA = AB/CA$, etc.

18. Let CE cut AB in T, AD cut CB in S, and the altitude from B cut AC in R. Then $AT/TB = EA/BC = AB/BC$, $BS/SC = AB/CD = AB/BC$, $CR/RA = (BC/AB)^2$.

19. Let t_A, t_B, t_C be the tangent lengths from D, E, F. Then $t_A^2 = (\overline{BD})(\overline{CD})$, $t_B^2 = (\overline{CE})(\overline{AE})$, $t_C^2 = (\overline{AF})(\overline{BF})$. But, by Menelaus' Theorem,

$$-1 = (\overline{BD}/\overline{DC})(\overline{CE}/\overline{EA})(\overline{AF}/\overline{FB}).$$

It now follows that $-(t_A t_B t_C)^2 = -[(AF)(BD)(CE)]^2$.

20. Multiply together results obtained by applying Menelaus' Theorem to triangles ABC, FBD, ECD, ECD, AFE, AFE, FBD with transversals DEF, AX, AX, BY, BY, CZ, CZ respectively.

21. Let AX, BY, CZ, cut BC, CA, AB in A', B', C'; let PO, QO, RO cut QR, RP, PQ in P', Q', R'. Then we have

$$(\overline{QP'}/\overline{P'R})(\overline{RQ'}/\overline{Q'P})(\overline{PR'}/\overline{R'Q}) = 1,$$

$$(\overline{QX}/\overline{XR})(\overline{RY}/\overline{YP})(\overline{PZ}/\overline{ZQ}) = 1,$$

$$(\overline{CP}/\overline{PB})(\overline{BR}/\overline{RA})(\overline{AQ}/\overline{QC}) = 1.$$

But, by Theorem 1.1.9,

$$(\overline{QP'}/\overline{P'R})/(\overline{QX}/\overline{XR}) = (\overline{CP}/\overline{PB})/(\overline{CA'}/\overline{A'B}),$$

$$(\overline{RQ'}/\overline{Q'P})/(\overline{RY}/\overline{YP}) = (\overline{AQ}/\overline{QC})/(\overline{AB'}/\overline{B'C}),$$

$$(\overline{PR'}/\overline{R'Q})/(\overline{PZ}/\overline{ZQ}) = (\overline{BR}/\overline{RA})/(\overline{BC'}/\overline{C'A}).$$

Setting the product of the left members of the above three equations equal to the product of the right members, it now follows that

$$(\overline{CA'}/\overline{A'B})(\overline{BC'}/\overline{C'A})(\overline{AB'}/\overline{B'C}) = 1.$$

22. We have $(\overline{B1}/\overline{1C})(\overline{CB'}/\overline{B'A})(\overline{A2}/\overline{2B}) = -1$, $(\overline{BA'}/\overline{A'C})(\overline{C3}/\overline{A3})(\overline{A2}/\overline{2B}) = -1$, $(\overline{AC'}/\overline{C'B})(\overline{B4}/\overline{4C})(\overline{C3}/\overline{3A}) = -1$, $(\overline{CB'}/\overline{B'A})(\overline{A5}/\overline{5B})(\overline{B4}/\overline{4C}) = -1$, $(\overline{BA'}/\overline{A'C})(\overline{C6}/\overline{6A})(\overline{A5}/\overline{5B}) = -1$, $(\overline{AC'}/\overline{C'B})(\overline{B7}/\overline{7C})(\overline{C6}/\overline{6A}) = -1$. Multiply together the first, fifth, and third of these equations; divide by the product of the other three; obtain $(\overline{B1}/\overline{C1})(\overline{C7}/\overline{B7}) = 1$, whence $1 \equiv 7$.

Section 1.5

3. On a line other than l through C construct A' and B' such that $\overline{CA'}/\overline{CB'} = r$. Let AA' and BB' intersect in D'. Through D' draw the parallel to CB' to intersect l in D.

4. (a) Expand.
   (b) Expand.

5. Insert an origin at O in the expansions of the two cross ratios. Now show

that $(\overline{OB} - \overline{OA})(\overline{OB}' - \overline{OC})(\overline{OA}' - \overline{OC}') =$
$(\overline{OB}' - \overline{OA}')(\overline{OB} - \overline{OC}')(\overline{OC} - \overline{OA})$.

6. By Theorem 1.5.2 (3), $(AC,BP) = 1 - m$, $(AC,BQ) = 1 - n$.

8. If A, B and C, D separate each other, then $(AB,CD) = r$, where $-\infty < r < 0$. Define θ such that $r = -\cot^2 \theta$. Let the semicircles on AB and CD intersect in V. Show that the angle of intersection of the two semicircles is 2θ.

Section 1.6

3. Let A', B', C', D', P' be the points of contact of the tangents a, b, c, d, p, respectively. Let O be the center of the circle and K any fixed point on the circle. Then $\measuredangle AOP' = (\measuredangle A'OP')/2 = \measuredangle A'KP'$, etc. It follows that $(AB,CD) = O(AB,CD) = K(A'B',C'D')$.

4. Let FA and DC intersect in T, BC and ED in U, BC and FE in V, and AF and DE in R. Consider the four tangents DE, AB, EF, CD. By Problem 3, Section 1.6, $(RA,FT) = (UB,VC)$, whence $D(EA,FC) = D(RA,FT) = E(UB,VC) = E(DB,FC)$. It now follows, by Corollary 1.6.3, that DA, EB, FC are concurrent.

5. Expand and use Menelaus' Theorem.

6. Set $(MN,AA') = k$. Expand and insert an origin at M. Solve for $\overline{MA}'$ to obtain $\overline{MA}' = (\overline{MA})(\overline{MN})/[k\overline{MN} + (1 - k)\overline{MA}]$, with similar expressions for $\overline{MB}'$, $\overline{MC}'$, $\overline{MD}'$. Now expand $(A'B',C'D')$; insert an origin at M; substitute the expressions for $\overline{MA}'$, $\overline{MB}'$, $\overline{MC}'$, $\overline{MD}'$; simplify to obtain (AB,CD).

7. The proof of Theorem 1.6.9 as given in the text applies here.

8. Let CC', $A'B$, $B'A$ be concurrent in P; let BB', $C'A$, $A'C$ be concurrent in Q. Let CB' and $B'C$ intersect in O, $PB'A$ cut $OC'B$ in X, PCC' cut QBB' in Y, and $PA'B$ cut $QC'A$ in Z. Then $A(OX,C'B) = B'(OX,C'Q) = B'(CP,C'Y) = Q(A'P,ZB) = A(A'X,C'B)$. It now follows that O, A, A' are collinear.

Section 1.7

2. Apply Theorem 1.7.3 (2) to the homographic pencils $A(X')$ and $B(X)$.

3. Apply Theorem 1.7.3 (2) to the homographic pencils $R(A)$ and $Q(A)$.

4. Apply Theorem 1.7.3 (1) to the homographic ranges (B) and (C).

5. (L) and (M) are homographic ranges. $F(L)$ and $H(M)$ are homographic pencils having FH as a common line.

6. $B(Y)$ and $C(X)$ are homographic pencils having BC as a common line.

7. (B) and (R) are homographic ranges having E as a common point.

8. Let Ω and Ω' be the infinite points on ranges (A) and (A') respectively. Then

$(AX,I\Omega) = (A'X',\Omega'J')$. That is, $\overline{AI}/\overline{IX} = \overline{J'X'}/\overline{A'J'}$. Therefore $(\overline{IX})(\overline{J'X'}) = (\overline{IA})(\overline{J'A'})$, a constant.

9. Let A and A' be any positions of X and X'. Then $(\overline{IX})(\overline{J'X'}) = (\overline{IA})(\overline{J'A'})$. Hence, reversing the steps in the proof of Problem 8, Section 1.7, we get $(AX,I\Omega) = (A'X',\Omega'J')$. Or, since $(\overline{IX})(\overline{J'X'})$ = constant, it immediately follows (from Theorem 1.7.4) that $(X) = (X')$. (This is the converse of Problem 8, Section 1.7.)

10. (a) Pencils $V(P)$ and $V(P')$ are congruent.
    (b) Apply Problem 8, Section 1.7.

11. Let the polygon be $A_1A_2 \ldots A_{n-1}A_n$, and let the sides A_1A_2, A_2A_3, $\ldots, A_{n-1}A_n$ each pass through a fixed point. Then $(A_1) = (A_n)$.

12. Parallel the proof of Problem 11, Section 1.7.

13. Use Theorem 1.6.7 and the facts that $AC/A'C' = VA/VC'$, $DB/D'B' = VB/VD'$, $CB/C'B' = VB/VC'$, $AD/A'D' = VA/VD'$.

Section 1.8

1. (a) $AC/CB = AP/MB = AP/BN = AD/DB$.
   (b) If O is the center of the semicircle, $(\overline{OC})(\overline{OD}) = (OT)^2 = (OB)^2$.
   (c) Apply Theorem 1.4.1.

2. $$(OB)^2 = (\overline{OC})(\overline{OD}) = (\overline{O'C} - \overline{O'O})(\overline{O'D} - \overline{O'O}) = (\overline{OO'} + \overline{O'C})(\overline{OO'} - \overline{O'C}).$$

3. (a) The angles in the pencil are multiples of 45°.
   (b) Let AB be the diameter, CD the chord, and P a point on the circle. Then $P(AB,CD) = A(TB,CD)$, where AT is the tangent at A. But $A(TB,CD) = (\sin \overline{TAC}/\sin \overline{CAB})/(\sin \overline{TAD}/\sin \overline{DAB})$.
   (c) $(AC,LM) = B(AC,LM) = B(AC,DE)$. Now use part (b).
   (d) Use part (b).

4. (a) Use Theorem 1.8.3.
   (b) Show that PA bisects $\measuredangle QPR$.
   (c) We have $AU/UT = [AB\cos(A/2)]/[BT\sin TBU]$ and also $AV/VT = [AC\cos(A/2)]/[TC\sin TCV]$, $AB/AC = BT/TC$, $\measuredangle TBU = \measuredangle TVC$.

5. Use Problem 1(b), Section 1.8.

6. By Theorem 1.1.9 show that $AC/CB = PA/PB = AD/DB$.

7. Use Theorem 1.8.11 and Problem 6, Section 1.8.

8. Draw any circle S cutting the circles drawn on AB and CD as diameters and let O be the intersection of the common chords of S and each of the other two circles. The circle with center O and radius equal to the tangent from O to either of the circles on AB and CD as diameters cuts the line $ABCD$ in the required points P and Q. The construction fails if the pairs A,B and C,D separate each other.

9. Use the theorems of Ceva and Menelaus.

10. $(TS)(TC)\sin ATB/(SA)(SC)\sin DSA = (\triangle ATC)/(\triangle ASC) = TU/SU = TV/SV = (\triangle BTD)/(\triangle BSD) = (TB)(TD)\sin ATB/(SB)(SD)\sin DSA$.

11. By Theorem 1.8.4, $2/\overline{AB} = 1/\overline{AC} + 1/\overline{AD} = (\overline{AD} + \overline{AC})/(\overline{AC}\cdot\overline{AD}) = 2\overline{AO}/(\overline{AC}\cdot\overline{AD})$.

12. $(\overline{OP})(\overline{OP'}) = (OB)^2 = (\overline{OQ})(\overline{OQ'})$.

13. Let AB cut PQ in O. Then O is the midpoint of PQ and $(\overline{OL})(\overline{OM}) = (\overline{OQ})^2$.

14. Let R and r be the radii of the semicircle and Σ respectively. Draw a concentric semicircle of radius $R - r$ and note that r is the geometric mean of $AC - r$ and $CB - r$.

15. Use Corollary 1.6.1.

16. We have $\overline{AC}/\overline{CB} = -\overline{AD}/\overline{DB}$, consequently $(AC)^2/(AD)^2 = (CB)^2/(DB)^2 = (\overline{OB} - \overline{OC})^2/(\overline{OB} - \overline{OD})^2 = (\overline{OD}\cdot\overline{OB} - \overline{OD}\cdot\overline{OC})^2/(OD)^2(\overline{OB} - \overline{OD})^2 = (\overline{OD}\cdot\overline{OB} - \overline{OB}\cdot\overline{OB})^2/(OD)^2(\overline{OB} - \overline{OD})^2 = (OB)^2/(OD)^2 = (\overline{OC})(\overline{OD})/(OD)^2 = \overline{OC}/\overline{OD}$.

17. We have $(DB, PV) = -1$, whence (by Problem 16, Section 1.8) $\overline{XP}/\overline{XV} = (DP)^2/(DV)^2$, with similar relations for Y and Z. Now apply Menelaus' Theorem to triangle PVU.

18. Draw the diametral secant AMN and let it cut the chord of contact $T'T$ of the tangents from A in R. Now, if O is the center of the given circle, $(\overline{OR})(\overline{OA}) = (OT)^2$. It follows that if circle Σ on AB as diameter cuts the given circle in P and Q, then OP is tangent to Σ. Thus, if S is the center of Σ, SP is tangent to the given circle. Therefore $(\overline{SC})(\overline{SD}) = (SP)^2 = (SB)^2$, and $(AB,CD) = -1$.

19. Take O, not on line ABC, and draw OA, OB, OC, OP, OQ, OR. Through A draw AM parallel to OQ, and let OB, OC, OP, OR cut AM in B', C', P', R'. Now B' is the midpoint of AC', P' is the first trisection point of $B'C'$, R' is the second trisection point of AB'. It follows that R' is the midpoint of AP', whence $O(QR,PA) = -1$.

20. The circle is a circle of Apollonius for PP' and for QQ'. It follows that if V is any point on this circle, then VB bisects angles PVP' and QVQ'.

22. Since $(AB,PQ) = -1$, we have $2/\overline{AB} = 1/\overline{AP} + 1/\overline{AQ}$. Therefore $2\overline{OB}/\overline{AB} = 2(\overline{OA} + \overline{AB})/\overline{AB} = 2\overline{OA}/\overline{AB} + 2 = (\overline{OA}/\overline{AP} + 1) + (\overline{OA}/\overline{AQ} + 1) = (\overline{OA} + \overline{AP})/\overline{AP} + (\overline{OA} + \overline{AQ})/\overline{AQ} = \overline{OP}/\overline{AP} + \overline{OQ}/\overline{AQ}$.

23. Take E and E' as a fixed pair of harmonic conjugates. Now ranges (F) and (F') are homographic. Therefore pencils $E(F)$ and $E'(F')$ are homographic. But EE' is a common line of these two pencils. It follows that the intersections of EF and $E'F'$ lie on a line. But B and C are two of these intersections. Therefore EF and $E'F'$ intersect on line BC.

24. Let R be the harmonic conjugate of O with respect to P and Q. Then the locus of R is a straight line n through the intersection of the two given lines, and $2/\overline{OR} = 1/\overline{OP} + 1/\overline{OQ}$. Now draw line n' parallel to n and through the

midpoint R' of OR. Then $1/\overline{OR'} = 1/\overline{OP} + 1/\overline{OQ}$, and n' is the sought line.

25. Replace $1/\overline{OP_1} + 1/\overline{OP_2}$ by $1/\overline{OX_1}$, $1/\overline{OX_1} + 1/\overline{OP_3}$ by $1/\overline{OX_2}$, and so on, using Problem 24, Section 1.8.

Section 1.9

2. Use Theorem 1.8.5.

3. Let the circles have centers O and O' and intersect in P. The circles are orthogonal if and only if triangle OPO' is a right triangle—that is, if and only if $[(c/2)d]/2 = (rr')/2$.

4. Use Theorem 1.9.2 (3).

5. Let M be the center of the given circle and N the center of circle CPQ. Now $\measuredangle PCQ = \measuredangle PNM$ and $\measuredangle PAQ = \measuredangle PMN$. Therefore $\measuredangle PNM + \measuredangle PMN = \measuredangle PCQ + \measuredangle PAQ = 180° - \measuredangle CQA = 90°$. It follows that triangle NPM is a right triangle and the two circles are orthogonal.

6. This is a converse of Problem 5, Section 1.9, and may be established by reversing the proof of Problem 5.

7. Use Problem 5, Section 1.9.

8. By symmetry the center of the circle through the four points must be the midpoint of the line of centers of the two given circles.

9. Let R and S be the centers of circles AMO and BMO respectively. Then $\measuredangle ROM = (\measuredangle AOM)/2$ and $\measuredangle SOM = (\measuredangle BOM)/2$. Therefore $\measuredangle ROS = 90°$.

10. $(BH)(BE) = (BD)(BC)$ and $(CH)(CF) = (CD)(CB)$. Therefore $(BH)(BE) + (CH)(CF) = (BC)^2$.

Section 1.10

1. The tangents drawn from the radical center to the three circles are all equal in length.

3. Use Theorems 1.10.6 and 1.10.7.

4. The midpoint of a common tangent to two circles has equal powers with respect to the two circles.

5. P is the radical center of the three circles.

6. (a) The altitudes are common chords of pairs of the circles.
(b) Let A', B', C' be the feet of the altitudes and H the orthocenter. Then AA', BB', CC' are chords of the three circles and $(\overline{AH})(\overline{HA'}) = (\overline{BH})(\overline{HB'}) = (\overline{CH})(\overline{HC'})$.

7. (a) Draw the diameter of C_1 through the center of C_2.
(b) Draw the chord of C_1 perpendicular to the diameter of C_1 through P.
(c) Use part (b) and the fact that P has equal powers with respect to the three circles.

(d) If circle $M(m)$ bisects the given circles $A(a)$ and $B(b)$, then $(MA)^2 + a^2 = m^2 = (MB)^2 + b^2$, whence $(MA)^2 - (MB)^2 = b^2 - a^2$.
(e) The center of the given circle has equal powers with respect to all the bisecting circles.

8. Use Theorem 1.10.6 (1).

9. (a) The center of the sought circle is the radical center of the two given circles and the given point.
(b) If the two given circles are intersecting, use part (a) and Theorem 1.10.11.

10. The required circle is the radical circle of the three circles having the given points as centers and the given tangent lengths as radii.

11. See Art. 113 in Johnson's *Modern Geometry*, or Art. 471 in Altshiller-Court's *College Geometry* (listed in bibliography of Chapter 1).

12. (a) A circle concentric with the given circle.
(b) A circle whose center is midway between the centers of the given circles.
(c) A straight line parallel to the radical axis of the given circles. Use Problem 11, Section 1.10.
(d) A circle coaxial with the given circles. Use Problem 11, Section 1.10.

13. (a) Use Theorem 1.10.9 (3).
(b) Use Theorem 1.10.11.

14. The point of concurrence is the radical center of the given circle and any two circles of the coaxial pencil.

15. If the three circles are distinct, their radical axes must be concurrent.

Section 2.1

1. (a) Onto and one-to-one.
(b) Not onto; 3 is not the image of any element of A.
(c) Not onto; 2 is not the image of any element of A.
(d) Not onto; no even integer is the image of any element of A.
(e) Onto and one-to-one.
(f) Onto and one-to-one.

3. Let r' be any real number and let r be a real root of $x^3 - x = r'$. Then $r \to r'$, whence the mapping is onto. The mapping is not one-to-one since $0 \to 0$ and $1 \to 0$.

4. (b) $n \to 4n^2 + 12n + 9$, $n \to 2n^2 + 3$, $n \to n^4$, $n \to 4n + 9$, $n \to 4n^4 + 12n^2 + 9$, $n \to 4n^4 + 12n^2 + 9$.

7. $S = IS = (T^{-1}T)S = T^{-1}(TS) = T^{-1}I = T^{-1}$.

8. Since $(T_2T_1)(T_1^{-1}T_2^{-1}) = I$ it follows (by Theorem 2.1.11) that $T_1^{-1}T_2^{-1} = (T_2T_1)^{-1}$.

9. $T^{-1}T = I$. Therefore, by Theorem 2.1.11, $T = (T^{-1})^{-1}$.

11. $(T_3T_2T_1)^{-1} = [(T_3T_2)T_1]^{-1} = T_1^{-1}(T_3T_2)^{-1} = T_1^{-1}(T_2^{-1}T_3^{-1}) = T_1^{-1}T_2^{-1}T_3^{-1}$.

12. (a) $(STS^{-1})^{-1} = (S^{-1})^{-1}T^{-1}S^{-1} = ST^{-1}S^{-1}$.
    (b) $(ST)(TS)(ST)^{-1} = (ST)(ST)(T^{-1}S^{-1}) = ST$.
    (c) $(ST_2S^{-1})(ST_1S^{-1}) = S(T_2T_1)S^{-1}$.

13. G1 is satisfied by Definition 2.1.5, G2 by Theorem 2.1.9, G3 by Theorem 2.1.10 (1), and G4 by Theorem 2.1.10(2).

Section 2.2

1. (a) (2,1), (b) (−1,1), (c) (1, −1), (d) (1,−1), (e) (−1,−1), (f) (2,2), (g) (4,−2), (h) (1,1/2), (i) (2, −1), (j) (2,−2), (k) (−2,2), (l) $(2-2\sqrt{2},0)$.

2. No, for all parts.

3. If $A = A'$, then $O = A$. If $B = B'$, then $O = B$. If $A \neq A'$, $B \neq B'$, and AA' is not parallel to BB', then O is the point of intersection of the perpendicular bisectors of AA' and BB'. If $A \neq A'$, $B \neq B'$, AA' is parallel to BB', and AB is not collinear with $A'B'$, then O is the point of intersection of lines AB and $A'B'$. If AB is collinear with $A'B'$, then O is the common midpoint of AA' and BB'.

4. Obvious from a figure.

5. Place an x axis on BC with origin at the midpoint of BC.

8. Obvious from a figure.

9. (a), (b), (c) Obvious from a figure.

10. (a) The midpoints of the sides of a quadrilateral form a parallelogram.
    (b) Use part (a).

11. Obvious from a figure.

12. Let $R(O,\theta)$ carry P and Q into P'' and Q''; let $R(O',-\theta)$ carry P'' and Q'' into P' and Q'. Now the angle from directed line PQ to directed line $P''Q''$ is θ, and the angle from directed line $P''Q''$ to directed line $P'Q'$ is $-\theta$. It follows that $P'Q'$ is parallel to PQ. Also, $P'Q' = PQ$.

13. On line O_1O_2 such that $\overline{O_2P} = (1-k_1)\overline{O_1O_2}/(k_1-k_2)$.

14. Let $T(AB)$ carry P into P_1, $R(l_1)$ carry P_1 into P_2, $T(CD)$ carry P_2 into P_3, $R(l_2)$ carry P_3 into P', $R(l_2)$ carry P_2 into P_4. Then $T(CD)$ carries P_4 into P'. From the figure it is easily seen that $P_2P_3P'P_4$ is a rectangle and that $\measuredangle P_2OP_4 = 2\theta$.

Section 2.3

3. Let a common tangent to circles $A(a)$ and $B(b)$ cut the line of centers AB in point S. Draw the radii to the points of contact of the common tangent. Then, from similar triangles, $SA/SB = a/b$, etc.

5. Use Theorem 2.3.9.

6. Let P be any point on the circle of similitude of the two given circles $A(a)$, $B(b)$. Then $PA/PB = a/b$, or $(PA^2 - a^2)/(PB^2 - b^2) = a^2/b^2$. Now use Problem 12(d), Section 1.10. For a direct proof not depending upon Casey's Power Theorem, see Daus, *College Geometry*, p. 91 (listed in bibliography for Chapter 1).

7. Use the theorem of Menelaus.

8. This is a special case of Problem 7, Section 2.3.

9. Its distance from A is $k^2c/(k^2 - 1)$, where $k = a/b$.

10. Use Theorem 1.9.3.

11. (b) We have, power of $A_1 = (\overline{A_1M_2})(\overline{A_1H_2}) = (\overline{A_1A_3})(\overline{A_1H_2})/2$. Similarly, power of $A_1 = (\overline{A_1A_2})(\overline{A_1H_3})/2$. It follows that the power of A_1 is $[(\overline{A_1A_3})(\overline{A_1H_2}) + (\overline{A_1A_2})(\overline{A_1H_3})]/4$.

12. Let the given triangle ABC have altitudes AD, BE, CF, orthocenter H, circumcenter O, and nine-point center N. Let the tangential triangle $A'B'C'$ have circumcenter O'. Now triangles DEF and $A'B'C'$ are homothetic. But H is the incenter (or an excenter) of triangle DEF and O is the incenter (or corresponding excenter) of triangle $A'B'C'$. N is the circumcenter of triangle DEF and O' is the circumcenter of triangle $A'B'C'$. It follows that H, O, N, O' are collinear.

13. Let Y_1, Y_2, Y_3 be the images of X_1, X_2, X_3 under the homothety $H(H,2)$, which carries the nine-point circle into the circumcircle. Show that triangle $Y_1Y_2Y_3$ is equilateral.

14. If the Euler line is parallel to A_2A_3 we have $OM_1 = (A_1H_1)/3$, or $R\cos A_1 = (2R\sin A_2 \sin A_3)/3$, R the circumradius. That is, $\cos A_1/\sin A_2 \sin A_3 = 2/3$. But $\cos A_1 = -\cos(A_2 + A_3) = \sin A_2 \sin A_3 - \cos A_2 \cos A_3$, etc. Also see Problem E 259, *The American Mathematical Monthly*, Oct. 1937.

15. The trilinear polar of the orthocenter is the radical axis of the circumcircle and the nine-point circle.

Section 2.4

2. (a), (b) Use Theorem 2.4.7.

3. (a) Use Theorems 2.4.4 and 2.4.5.
   (b) Use Theorem 2.4.9.
   (c) Use Theorems 2.4.7 and 2.4.9.

4. (a) Obvious from a figure.

5. Use Theorem 2.4.9 and Lemma 2.4.8.

6. Obvious from a figure.

7. Easily shown from a figure.

8. Use Lemma 2.4.8.

9. (a) Use Theorem 2.4.9.
   (b) For brevity represent $R(l_1)$, $R(l_2)$, $R(l_3)$, by R_1, R_2, R_3 respectively. By Theorem 2.4.10, $R_1R_3R_2$ and $R_3R_2R_1$ are glide-reflections, whence $(R_1R_3R_2)^2$ and $(R_3R_2R_1)^2$ are translations. Since translations commute we have

$$(R_1R_3R_2)^2(R_3R_2R_1)^2 = (R_3R_2R_1)^2(R_1R_3R_2)^2,$$

or

$$R_1R_3R_2R_1R_3R_2R_3R_2R_1R_3R_2R_1 = R_3R_2R_1R_3R_2R_1R_1R_3R_2R_1R_3R_2,$$

or, since $R_1R_1 = I$,

$$R_1(R_3R_2R_1R_3R_2)^2R_1 = (R_3R_2R_1R_3R_2)^2,$$

or

$$R_1(R_3R_2R_1R_3R_2)^2 = (R_3R_2R_1R_3R_2)^2R_1.$$

Since R_1 and $(R_3R_2R_1R_3R_2)^2$ commute, it follows that $(R_3R_2R_1R_3R_2)^2$, which must be a translation or a rotation, is a translation along l_1.

(c)
$$\begin{aligned} G(l_3,A_1A_2)G(l_2,A_3A_1)G(l_1,A_2A_3) &= R_3T(A_1A_2)R_2T(A_3A_1)R_1T(A_2A_3) \\ &= R_3T(A_1A_2)T(A_3A_1)R_2R_1T(A_2A_3) \\ &= R_3T(A_3A_2)R_2R_1T(A_2A_3) \\ &= R_3T(A_3A_2)R(A_3,2\sphericalangle A_3)T(A_2A_3) \\ &= R_3R(A_2,2\sphericalangle A_3). \end{aligned}$$

(d) Use Theorem 2.4.5.
(e) $R_3\ R_2\ R_1$, being an opposite isometry, is a glide-reflection. To find which glide-reflection it is, consider the fates of H_1 and H_3.

10. $R(l_3)R(l_2)R(l_1)$ is an opposite isometry, and therefore (by Theorem 2.4.9) a reflection in a line or a glide-reflection. But O is an invariant point.

Section 2.5

2. (a), (b) Use Theorems 2.5.3 and 2.5.4.

3. (a) Use Theorem 2.4.9.
   (b) Use Theorem 2.5.4.

4. (a), (b) Obvious from a figure.

5. If the isometry is opposite, it is a reflection or a glide-reflection; in either case the midpoints of segments PP' lie on the axis of the transformation. If the isometry is direct, the result follows from Theorem 2.5.6, the locus degenerating to a point if the isometry is a half-turn.

6. The maps are related by a direct nonisometric similarity, that is, by a homology.

7. (a) A translation or a homology.
   (b) A glide-reflection or a stretch-reflection.

8. Consider a triangle ABC; it is carried into a triangle $A'B'C'$ whose sides are

parallel to the corresponding sides of triangle ABC. It follows that the two triangles are coaxial (on the line at infinity), and thus (by Desargues' two-triangle theorem) also copolar at a point P. If P is an ideal point, the similarity is the translation $T(AA')$; if P is an ordinary point, the similarity is the homothety $H(P,\overline{A'B'}/\overline{AB})$.

10. (a), (b) Use Theorem 2.5.6.

Section 2.6

1. (a) The figure consists of certain parts of four concurrent circles.
   (b) The figure consists of certain parts of two circles and of two lines.

2. (a) A system of coaxial intersecting circles.
   (b) A system of coaxial tangent circles.

4. (a) Let OB cut the given circle again in M. Then $(\overline{OM})(\overline{OB}) = (\overline{OC})(\overline{OC}')$, or $\overline{OB} = (\overline{OC})(\overline{OC}')/2(\overline{OC}) = \overline{OC}'/2$.
   (b) Denote the center of inversion by O and the center of K' by M, and let OM cut K in S and T and K' in S' and T', where S and S', and T and T', are corresponding points under the inversion. Now $(\overline{OC})(\overline{OC}') = (\overline{OS})(\overline{OS}') = (\overline{OT})(\overline{OT}')$ and

$$\begin{aligned}(\overline{MO})(\overline{MC}') &= \overline{OM}(\overline{OM} - \overline{OC}') = \overline{OM}[(\overline{OS}' + \overline{OT}')/2 - (\overline{OT})(\overline{OT}')/\overline{OC}] \\ &= \overline{OM}[(\overline{OS}' + \overline{OT}')/2 - 2(\overline{OT})(\overline{OT}')/(\overline{OS} + \overline{OT})] \\ &= \overline{OM}[(\overline{OS}' + \overline{OT}')(\overline{OS} + \overline{OT}) - 4(\overline{OT})(\overline{OT}')]/2(\overline{OS} + \overline{OT}) \\ &= \overline{OM}(\overline{OT} - \overline{OS})(\overline{OS}' - \overline{OT}')/4\overline{OC} = \overline{OM}(2\overline{CT})(2\overline{MS}')/4\overline{OC} \\ &= (\overline{MS}'/\overline{CT})(\overline{CT})(\overline{MS}') = (MS')^2.\end{aligned}$$

   There are much easier ways to solve this problem; see, e.g., Problem 12, Section 2.7.
   (d) Let the centers of the circles be C and C' and let their common chord ST cut CC' in M. Then $(\overline{CM})(\overline{CC}') = (CS)^2$.

Section 2.7

1. Invert with respect to O.

2. Invert with respect to A.

3. (a) If C' is any point on a semicircle having diameter $A'OB'$, the circles $A'OC'$ and $B'OC'$ are orthogonal.
   (b) If B', C', D' are three collinear points and A is a point not collinear with them, the circles $AB'D'$ and $AC'D'$ intersect the lines AB' and AD', respectively, in equal or supplementary angles.

4. Invert with respect to A.

5. Invert with respect to M.

6. Invert with respect to any one of the four points.

7. Invert with respect to T.

8. Invert with respect to a point of intersection of K and K_2.

9. Invert with respect to A.

10. Invert with respect to the given point and discover that the sought locus is the circle through the given point and orthogonal to the system of coaxial circles; its center lies on the common tangent of the system and passes through the common point of the system.

11. Invert with respect to C.

Section 2.8

1. Consider two circles orthogonal to C_1 and C_2.

2. Invert with respect to O, then use Menelaus' Theorem and Theorem 2.7.5.

3. Invert with respect to a point on C, and show that $t_{ij}^2/r_i r_j$ is invariant.

4. (a) Let P be any point. Let P' be the inverse of P for circle $A(a)$ and let P'' be the inverse of P' for circle $B(b)$. Let Q be the inverse of P for circle $B(b)$. Now invert the figure with respect to circle $B(b)$. Since circle $A(a)$ inverts into itself (Theorem 2.6.4) and P and P' are inverse points for circle $A(a)$, it follows (by Theorem 2.7.4) that Q and P'' are also inverse points for circle $A(a)$.
(b) Invert the figure in circle K_2 and use Theorem 2.7.4.

5. Invert with respect to D. Then $A'C' + C'B' = A'B'$. But, if p' is the perpendicular distance from D to line $A'B'C'$, we have $A'C'/p' = AC/r$. Similarly, $C'B'/p' = CB/q$, $A'B'/p' = AB/p$, etc.

6. Let a secant through O cut C in A and B and cut C' in A' and B', where A,A' and B,B' are pairs of inverse points. Then $pp' = (\overline{OA})(\overline{OB})(\overline{OA'})(\overline{OB'}) = r^4$.

7. Invert the coaxial system into a system of concentric circles, a system of concurrent lines, or a system of parallel lines.

8. (b) Let A and A' be a pair of antinverse points for circle K and let J be any circle through A and A'. Let O be the center of K and let P and Q be the points of intersection of K and J. Let PO cut J again in R. Then $(\overline{OP})(\overline{OR}) = (\overline{OA})(\overline{OA'}) = -r^2$, where r is the radius of K. It follows that $P \equiv Q$.

9. Let triangles PQR and UVW, inscribed in a circle K, be copolar at C. Let U', V', W' be the inverse of U, V, W for C as center of inversion and r^2 as power of inversion. Let p denote the power of C with respect to circle K. Then $\overline{CU'}/\overline{CP} = r^2/(\overline{CU})(\overline{CP}) = r^2/p$.

10. (a) Such a point is the center of a circle orthogonal to both of the given circles. Invert with respect to this circle.
(b) When the radical center is outside all three circles.

11. Use Theorem 2.7.7.

12. Invert with respect to any point on their radical circle.

13. Use Theorem 2.7.5 and some circles of Apollonius.

14. Use Problem 13, Section 2.8.

15. Use Theorem 2.7.4 (2).

16. Use Ptolemy's Theorem.

17. Subject the figure to the inversion $I(A,1)$.

18. Invert with respect to D and then apply Stewart's Theorem (see Problem 13 of Section 2.1).

19. (b) Let C_1 be the circle orthogonal to circle $ABCD$ and passing through A and C; let C_2 be the circle orthogonal to circle $ABCD$ and passing through B and D. Let C_1 and C_2 intersect in X and Y. Invert with respect to X.

22. We have $OP = (BD)(AO)/AB$ and $OP' = (AC)(OB)/AB$, whence $(OP)(OP') = (BD)(AC)(AO)(OB)/(AB)^2$. But $(BD)(AC) = (AD)^2 - (AB)^2$.

23. If V is outside the circle, invert the circle into itself with respect to V as center of inversion. If V is inside the circle, invert the circle into its reflection in V.

Section 2.9

3. Let O and r be the center and radius of the circle and let Q' be the inverse of Q. Then $(PQ)^2 = (OP)^2 + (OQ)^2 - 2(OQ)(OQ') = (OP)^2 + (OQ)^2 - 2r^2$.

4. By Theorem 2.6.7, P and Q are inverse points for circle K_2.

5. (a), (b) Use Theorems 2.9.7, 2.6.7, and 1.9.3.

6. (a) Let Q be diametrically opposite P on circle R. Then (by Problem 5(b), Section 2.9) P, Q are conjugate points for K_1, K_2, K_3. Therefore the poles of P for K_1, K_2, K_3 all pass through Q.
   (b) Draw the circle on PQ as diameter. Prove that this circle is orthogonal to the coaxial system. Then use Problem 5(b), Section 2.9.
   (c) Use part (b).
   (d) Use part (b).
   (e) Use Problem 5(a) and (b), Section 2.9.

7. Let P', Q' be the inverses of P, Q. Show that $OP'YQ$ is similar to $OQ'XP$.

Section 2.10

1. Line t is the polar of point T.

2. Let T be the point of contact of the tangent to the incircle. Then the poles, for the incircle, of AP, BQ, CR are the feet of the perpendiculars from T on the sides of the triangle determined by the points of contact of the incircle with the sides of triangle ABC.

7. (a) Use Problem 6, Section 2.10.

(b) Use Problem 6, Section 2.10.
(c) The inverse of each vertex of the triangle is the foot of the altitude through that vertex.

8. (a) Let AB and CD intersect in M and let the tangents to the circle at A and B intersect in N. Then $-1 = (NM,CD) = B(NM,CD) = B(BA,CD) = (BA,CD) = (AB,CD)$.
(b) By Theorem 1.6.7, $(AB,CD) = e(AC/CB)/(AD/DB)$. Therefore $(AC)(BD) = (BC)(AD)$. The rest follows by Ptolemy's Theorem.

9. (a) Use Theorem 2.10.2.
(b) Use part (a) and Problem 5(a), Section 2.9.

11. (a) The pole, with respect to the incircle, of YZ is A. Let P' be the pole of AA'. Then, since YZ and AA' are conjugate lines, $A(BC,A'P') = -1$, whence AP' is parallel to BC. It now follows that the poles of AA', PX, YZ are collinear.
(b) Let the line through P parallel to BC cut YZ in S. Since YZ and PM are conjugate lines, $P(ZY,MS) = -1$. Therefore $(RQ,M\infty) = -1$ and M is the midpoint of RQ.

Section 2.11

2. (a), (b), (c) Yes.

4. There is a unique isometry that carries a given noncoplanar tetrad of points A, B, C, D into a given congruent tetrad A', B', C', D'.

5. (a) See the proof of Theorem 2.4.6.
(b) See the proof of Theorem 2.4.5.

7. Use Theorem 2.11.4.

8. Use Problem 7, Section 2.11.

11. Generalize the proof of Theorem 2.6.4.

13. For the first part, generalize the proofs of Theorems 2.6.10, 2.6.11, 2.6.12, 2.6.13. The second part follows since the intersection of two "spheres" is a "circle."

15. Invert with respect to the point of contact of S_1 and S_2.

16. (a), (b), (c) Generalize the proof of Theorem 2.9.3.

Section 3.1

2. (b) Let A be the given point and BC the given line segment. Construct, by Proposition 1, an equilateral triangle ABD. Draw circle $B(C)$, and let DB produced cut this circle in G. Now draw circle $D(G)$ to cut DA produced in L. Then AL is the sought segment.

3. It is a matter of existence; there exists a greatest triangle inscribed in a circle, but there does not exist a greatest natural number. To complete argument I,

we must prove that a maximum triangle inscribed in a circle *exists*. The problem illustrates the importance in mathematics of existence theorems.

Section 3.2

1. (8) Let A and B be the two given points, P any point on the locus, and M the foot of the perpendicular from P on AB. Then $(MA)^2 - (MB)^2 = [(PM)^2 + (MA)^2] - [(PM)^2 + (MB)^2] = (PA)^2 - (PB)^2 =$ a constant, whence M is fixed. (9) Let A and B be the two given points, P any point on the locus, and O the midpoint of AB. Then, by the law of cosines,

$$(PA)^2 = (PO)^2 + (AO)^2 - 2(PO)(AO)\cos(AOP),$$
$$(PB)^2 = (PO)^2 + (BO)^2 - 2(PO)(BO)\cos(BOP).$$

   It follows that $(PA)^2 + (PB)^2 = 2(PO)^2 + (AB)^2/2 =$ a constant, whence $PO =$ constant.

3. Use loci (1) and (2).

4. Let r be half the distance between the two parallel lines. Draw the circle of radius r with center at the given point, to intersect the line midway between the two parallel lines in the centers of the sought circles.

5. Use loci (5) and (1).

6. Locate the point as an intersection of some circles of Apollonius.

7. Denote the initial positions of the balls by A and B, and the center of the table by O. Construct the circle of Apollonius of A and B for the ratio $\overline{AO}/\overline{OB}$.

8. Let the given points be A and B and let the required chords be AC and BD. Let E, F be the midpoints of AB, CD; let O be the center of the circle. Then F lies on $O(E)$ and on $E(s/2)$, where $s = AC + BD$.

9. Let R and r be the radii of the given and required circles. Draw circles of radii $R \pm r$ concentric with the given circle.

10. Locate the vertex of the right angle by the method of loci.

11. Use loci (1) and (2).

12. Draw a line parallel to the given line at a distance from the given line equal to the radius of the given circle. Let P be the foot of the perpendicular from the given point to the drawn line. Find the center of the required circle as the intersection of the perpendicular to the given line at the given point, and the perpendicular bisector of PC, where C is the center of the given circle.

13. Draw a circle concentric with the given circle and having a radius equal to the hypotenuse of a right triangle having legs equal to the tangent length and the radius of the given circle.

14. In cyclic quadrilateral $ABCD$, let us be given angle A, AB, BD, AC. First construct triangle ABD and then find C by loci (1) and (6).

15. Locate the midpoint of the chord cut off by the circle on the required line by loci (2) and (5).

16. Use two circles of Apollonius.

17. It is a rectangle having the given lines as diagonal lines.

18. (a) Use the locus of Problem 17, Section 3.2.
(b) Draw a rectangle having the given lines as diagonal lines and having one side tangent to the given circle.

Section 3.3

1. Translate one circle through a vector determined by the given line segment.

2. Use Problem 1, Section 3.3.

3. Reflect two adjacent sides of the quadrilateral in the given point.

4. Reflect one of the curves in the given line.

5. Use Problem 4, Section 3.3.

6. Let ABC be the sought triangle and let D, E, F be the given points on the sides BC, CA, AB respectively. Let $\overline{BD}/\overline{DC} = m/n$, $\overline{CE}/\overline{EA} = p/q$, $\overline{AF}/\overline{FB} = r/s$. Find F_1 on line FE such that $\overline{FE}/\overline{EF_1} = q/p$. Next find F_2 on line F_1D such that $\overline{F_1D}/\overline{DF_2} = n/m$. Then F_2F lies along the line AB, etc.

7. Use the method of similitude.

8. Let the given line and the given directrix meet in a point O; draw the line determined by O and the given focus F. Take any point on the given line as center and draw a circle tangent to the directrix. Now use O as a center of homothety.

9. Use the method of similitude.

10. Take any point D' on BA. Then take E'' on CA such that $CE'' = BD'$. Let circle $D'(B)$ cut the parallel to BC through E'' in E'. Draw a line through E' parallel to AC to cut BA in A' and BC in C'. We now have a figure homothetic to the desired figure, with B as center of homothety.

11. Use Problem 10, Section 3.3.

12. Subject C_2 to the homothety $H(O,k)$, where $k = OP_1/OP_2$.

13. Subject the given circle to the homothety $H(O,-k)$, where k is the given ratio.

14. Use Problem 12, Section 3.3.

15. Use the method of similitude.

16. Use the method of similitude.

17. Use General Problem 3.3.4.

18. Take any point O on one of the circles and let C_1 and C_2 denote the other two circles. Now use General Problem 3.3.4.

19. Let $(OP_1)(OP_2) = a^2$. Invert C_2 into C_2' in the circle $O(a)$.

20. Use Problem 19, Section 3.3.

21. Use Problem 19, Section 3.3.

22. Invert with respect to the given point.

23. This is a special case of Problem 22, Section 3.3.

24. Increase the radius of each circle by half the distance between two of them to obtain three circles, two of which are externally tangent to one another. Invert with respect to the point of contact of these two circles.

Section 3.4

1. Let A be an arbitrary point on m and let A' be the corresponding point on m'. Let PA cut m in A''. Then range (A'') is homographic to range (A'). If D is a double point of these two coaxial homographic ranges, PD is a desired line.

2. Find the double points of the homographic ranges cut by the two given pencils on line m.

3. Join the vertices V and V' of the pencils, and on VV' describe the circular arc containing the given angle. Let corresponding rays of the two homographic pencils cut this arc in points P and P'. We seek the double points of the concyclic homographic ranges (P) and (P').

4. Draw a line m parallel to the join of the vertices V and V' of the given pencils. Let corresponding lines of the two pencils cut m in A and A'. Mark off A'' on m such that $\overline{AA''} = \overline{VV'}$. Find the common points of the two coaxial homographic ranges (A') and (A'').

5. Let the two given lines be m and m' and the two given points be V and V'. Let A be any point on m and find A' and A'' on m' such that angles AVA' and $AV'A''$ are equal to the given angles. Now find the double points of the two coaxial homographic ranges (A') and (A'').

6. Let ABC be the given triangle, m the given line, and P any point on m. Let P' and P'' be points on m such that AP and AP' are isogonal lines for vertex A and BP and BP'' are isogonal lines for vertex B. Now find the double points of the two coaxial homographic ranges (P') and (P'').

7. Take any point A' on BC, draw $A'C'$ in the assigned direction to cut BA in C', draw $C'B'$ in the assigned direction to cut AC in B'. Draw $A'B'$ in the assigned direction to cut AC in B''. Now find the double points of the two coaxial homographic ranges (B') and (B'').

8. Take any point A on p and find the feet R and S of the perpendiculars from A on r and s. Find R' and S' on r and s such that RR' and SS' have the given

projected lengths. Let the perpendicular to r at R' and the perpendicular to s at S' cut q in B and B' respectively. Now find the double points of the two coaxial homographic ranges (B) and (B').

9. Let P be the given point and m and n the given lines. Let A be any point on m and let PA cut n in A'. Mark off the given lengths AB and $A'B''$ on m and n respectively, and let PB cut n in B'. We seek the double points of the coaxial homographic ranges (B') and (B'').

10. Take any point A on m and let AP cut m' in A'. Find A'' on m' such that $OA/O'A''$ is the given constant. Now find the double points of the two coaxial homographic ranges (A') and (A'').

11. Parallel the solution suggested for Problem 10, Section 3.4.

12. Let m and m' be the two given lines, O their point of intersection, and P the given point. Take any point A on m and let AP cut m' in A'. Find A'' on m' such that triangle AOA'' has the given area. Now find the double points of the two coaxial homographic ranges (A') and (A'').

13. Let ABC be the given triangle and P, Q, R the given points. Let A' be any point on BC. Find B' on CA such that $\sphericalangle A'RB'$ is equal to the first of the given angles; next find C' on AB such that $\sphericalangle B'PC'$ is equal to the second of the given angles; then find A'' on BC such that $\sphericalangle C'QA''$ is equal to the third of the given angles. We seek the double points of the coaxial homographic ranges (A') and (A'').

14. Let P, Q, R be the three given points and let A be any point on the circle. Draw AP to cut the circle again in B; next draw BQ to cut the circle again in C; then draw CR to cut the circle again in A'. We seek the double points of the concyclic homographic ranges (A) and (A'). See Problems 13, Section 1.7, and 23, Section 2.8.

15. As in Problem 14, Section 3.4, inscribe in the circle a triangle whose sides shall pass through the poles of the given lines, and then draw the tangents to the circle at the vertices of this triangle.

Section 3.5

1. (b) Use successive applications of part (a).
   (d) Case 2. Find N on OM such that $ON = n(OM) > (OD)/2$. By Case 1 find N', the inverse of N in $O(D)$. Finally find M' such that $OM' = n(ON')$.
   (e) See Problem 4 (a), Section 2.6.
   (f) See Problem 4 (b), Section 2.6.
   (g) From the points A, B, C, D one can, with a Euclidean compass alone, obtain a circle k whose center is not on AB or $C(D)$. Under inversion in k, line AB and circle $C(D)$ become circles whose centers are constructible (by parts (e) and (f)), and points on which are constructible (by part (d)). These circles can then be drawn, and their intersections found. The inverses in k of these intersections are the sought points X and Y.
   (h) From the points A, B, C, D one can, with a Euclidean compass alone, obtain a circle k whose center is not on AB or CD. Under inversion in k,

lines AB and CD become circles through the center O of inversion. The centers of these circles are constructible (by parts (e) and (f)), and, since they pass through O, the circles can be drawn. The inverse in k of the other point of intersection of these circles is the sought point X.

2. Circle ABC is the inverse of line BC' in circle $A(B)$. Hence use Problem 4 (a), Section 2.6.

3. Find C such that $\overline{AC} = 2\overline{AB}$. Let $A(B)$ and $C(A)$ intersect in X and Y. Draw $X(A)$ and $Y(A)$ to intersect in the sought midpoint M.

4. We suppose the center O of the circle is given. Draw $A(C)$ and $D(B)$ to intersect in M. Draw $A(OM)$ to cut the given circle in X, Y. Then A, X, D, Y are vertices of an inscribed square. The proof is easy.

Section 3.6

2. Case 1, P not on k. Draw PAB, PCD cutting k in A, B and C, D. Draw AD, BC to intersect in M. Draw AC, BD to intersect in N. Then MN is the sought polar.
   Case 2, P on k. Draw any secant m through P and let R and S be any two points on m but not on k. Find the polars r and s of R and S. Then r and s intersect in M, the pole of m. PM is the sought polar.

3. (a) Find, by Problem 2, Section 3.6, the polar p of P, and let p cut k in S and T. Then PS and PT are the sought tangents.
   (b) Find the polar p of P by Problem 2, Section 3.6. Or inscribe a hexagon 123456 in k, where $1 \equiv 2 \equiv P$, and use Pascal's mystic hexagram theorem.

4. (a) Draw (by Problem 3.6.1) a line m parallel to line ABC. Choose a point V not on m or ABC and let VA, VB, VC cut m in A', B', C'. Let BA' and CB' intersect in V'. Draw $V'C'$ to cut ABC in D; draw DV to cut m in D'; draw $V'D'$ to cut ABC in E; etc.
   (b) We illustrate with $n = 5$. In the figure of the solution for part (a), let $A'A$ and $F'B$ intersect in U. Then $B'U$ cuts AB in a point Y such that $AY = AB/5$.

Section 3.7

1. Consider an oblique cone with a circular base and vertex V. There exist circular sections of the cone that are not parallel to the base; let c be one of these circular sections. A straightedge construction of the center of the circular base of the cone would lead, by projection from V, to a straightedge construction of the center of c. But the center of c is not the projection from V of the center of the base.

2. (a) By Problem 3.6.1 draw through P lines parallel to the diagonals of the given parallelogram to cut AB in C and D. By Problem 3.6.1 draw CE parallel to PD and DE parallel to PC, and let PE cut CD in M. Now, by Problem 3.6.1, draw through P a line parallel to AB.
   (b) By part (a) draw three chords parallel to one diagonal of the parallelogram. Now, by connecting opposite ends of pairs of these chords, obtain two points on the diameter of k that is perpendicular to the three chords; draw

this diameter. Carry out a similar construction with respect to the other diagonal of the parallelogram.

3. (a) Take M and N, any two points on the inner circle and such that MN is not a diameter of the circle. By Problem 3 (b), Section 3.6, draw the tangents to the inner circle at M and N. We now have, in the outer circle, two bisected chords in two different directions. This allows us (by Problem 3.6.1) to construct a parallelogram with sides parallel to these bisected chords. Now use Problem 2 (b), Section 3.7.
(b) Let P be the point of contact. Draw any three secants AA', BB', CC' through P, where A, B, C are on one circle and A', B', C' are on the other. Then BA is parallel to $B'A'$, and BC is parallel to $B'C'$. Now use Problem 2 (b), Section 3.7.
(c) Let the two circles intersect in P and Q. Draw any two secants MQT and UQS where M and U are on one circle and T and S are on the other. Next draw secants TPV, SPN, where V and N are on the other circle. Then MN is parallel to VU. Similarly obtain another pair of parallel lines, and use Problem 2 (b), Section 3.7.

4. Draw any two secant lines through the center of similitude. Connect pairs of corresponding points of intersection of these secant lines with the circles to obtain two pairs of parallel lines. Now use Problem 2 (b), Section 3.7.

5. (a) Take any point P on AB and with the parallel ruler draw any pair of parallel lines PP', MM' the width of the ruler apart. Let MM' cut AB in M. Then draw NN' parallel to PP' and the width of the ruler from PP'. Let NN' cut AB in N. Then P is the midpoint of segment MN.
(b) Use Problem 3.6.1.
(c) Suppose P is on AB. Draw any line $P'P$ through P and then draw $M'M$ and $N'N$ parallel to $P'P$ the width of the ruler from $P'P$. Then draw NN'' not parallel to $P'P$ and the width of the ruler from P. Let NN'' meet $M'M$ in R. RP is the required perpendicular.
Suppose P is not on AB. Take any point Q on AB and (by the above case) draw QR perpendicular to AB. By part (b) draw PT parallel to RQ. Then PT is the required perpendicular.
(d) Let k denote the width of the ruler. Take any point C' and draw $C'D'$ parallel to CD. Take D' such that $C'D' = k$; this can be done by erecting a perpendicular to $C'D'$ at C', and then using the ruler to obtain D'. Draw CC' and DD' to meet in S. Draw $C'A'$ parallel to CA to meet SA in A'. Then draw $A'B'$ parallel to AB. We have now reduced the problem to that where $CD = k$. We proceed, then, with this special case.
Take any point P on AB outside the circle $C(D)$. Draw PT at distance k from C; PT is then a tangent to $C(D)$. Draw CT perpendicular to PT. Draw CR perpendicular to AB, TR perpendicular to PC. Through R draw RX and RY at distance k from C to cut AB in the required points X and Y. The proof is easy. TR is the polar of P, hence AB is the polar of R. Hence, since RX and RY are tangent to $C(D)$, X and Y are the required points of intersection of AB and $C(D)$.
(e) We may proceed exactly as in Problems 3.6.5 and 3.6.6.

9. Proof I. Let C be a circle inside S and such that G', A', R' lie outside C.

Invert G', A' in C, obtaining G'', A'' inside C. Find R'' (also inside C) and then invert R'' in C to obtain R'.
PROOF II. Take point P inside S as the center of homothety and construct, from G', A', loci G'', A'' homothetic to G, A, but lying entirely inside S. Now construct R'' (also inside S) and then obtain R'.

12. (a) Use the fact that the sum of the infinite geometric series $\frac{1}{2} - \frac{1}{4} + \frac{1}{8} - \frac{1}{16} + \cdots$ is $\frac{1}{3}$. For another asymptotic Euclidean solution of the trisection problem see Problem 4134, *The American Mathematical Monthly*, Dec. 1945.

15. (a) Note that $\sphericalangle OLP = \sphericalangle ORL + \sphericalangle LOR$ and $\sphericalangle OPQ = \sphericalangle OPL + \sphericalangle LPQ$. But $\sphericalangle OLP = \sphericalangle OPQ$ and $\sphericalangle ORL = \sphericalangle OPL$. Therefore $\sphericalangle LOR = \sphericalangle LPQ = \sphericalangle LOQ = 30°$ and $\sphericalangle LKQ = 2(\sphericalangle LOQ) = 60°$. It follows that $RL = QL = KL$.

Section 3.8

14. Show that $\tan(180°/17)$ is approximately equal to $\frac{3}{16}$.

Section 3.9

1. If a regular polygon of nine sides can be constructed, then its central angle of 40° can be constructed, whence a 60° angle can be trisected.

2. If an angle of 1° can be constructed, then so also can an angle of 20°, whence a 60° angle can be trisected.

3. Let $7\theta = 360°$. Then $\cos 3\theta = \cos 4\theta$ or, if we set $x = \cos\theta$, $8x^3 + 4x^2 - 4x - 1 = 0$.

5. Use the identity of Problem 4.

6. Take the diameter of the circle as 1 unit. Then the circumference equals π.

8. Show that BB' and AA' are two mean proportionals between OA and OB. If $OB = 2(OA)$, then $(BB')^3 = 2(OA)^3$. This solution of the duplication problem was given by Apollonius (*ca.* 225 B.C.).

9. Show that $(AC)^3 = 2(AB)^3$. Essentially this construction for duplicating the cube was given in publications by Viète (1646) and Newton (1728).

10. Show that angle ADB = (angle AOB)/3. This solution of the trisection problem is implied by a theorem given by Archimedes (*ca.* 240 B.C.).

11. Take $AOB = 90°$, and let M and N be the feet of the perpendiculars from P on OA and OB. Let R be the center of the rectangle $OMPN$. If CD is Philon's line for angle AOB and point P, then $RE = RP$, and hence $RD = RC$. We now have a solution of Problem 8.

Appendix

2. (a) The sides of the angle that the vanishing points of m and n subtend at V are parallel to m' and n'.
(b) Use part (a).

3. (a) Denote the required center of perspectivity by V. Then (by Problem 2 (a), above) $\sphericalangle AVC = \sphericalangle A'B'C' = \alpha$ and $\sphericalangle DVF = \sphericalangle D'E'F' = \beta$. To find V, draw on AC an arc of a circle containing angle α and on DF (on the same side of l) an arc of a circle containing angle β. Since A, C, D, F are in the order A, D, C, F, these arcs must intersect; let X be such an intersection. Now rotate X about line $ADCF$ out of plane π to a position V. Then if we project plane π from center V onto a plane parallel to the plane of V and l, the problem is solved.
   (b) Not necessarily—only so long as the circular arcs described in the solution of part (a) intersect one another.

4. Let AB and CD intersect in U, AD and BC in V, AC and BD in W. By Problem 3 above, project line UV to infinity and angles VAU and LWM into right angles.

5. Draw any line cutting the rays of the pencil $U(AB,CD)$ in (ab,cd). Let $U'(A'B',C'D')$ be the projection, under any perspectivity, of the pencil $U(AB,CD)$, and let $(a'b',c'd')$ be the projection of (ab,cd). Then $U(AB,CD) - (ab,cd) = (a'b',c'd') = U'(A'B',C'D')$.

6. Let V be the center of perspectivity, and let X and Y' be the feet of the perpendiculars from V on π and π'. Then the bisectors of $\sphericalangle XVY'$ cut π and π' in the isocenters of the perspectivity. Suppose, for example, the internal bisector of $\sphericalangle XVY'$ cuts π and π' in E and E'. Then XE and $E'Y'$ intersect on the axis of perspectivity in a point K, and $EK = E'K$. Let the sides of an angle at E cut the axis of perspectivity in L and M. Then $\sphericalangle LEM$ maps into $\sphericalangle LE'M$. But triangles LEM and $LE'M$ are congruent, etc.

7. The isolines are the reflections of the axis of perspectivity in the vanishing lines of the two planes.

Index